Conservation of Bridges

Conservation of Bridges

Graham Tilly, Gifford and Partners

in association with
Alan Frost, Donald Insall Associates
and
Jon Wallsgrove, The Highways Agency

First published 2002 by Spon Press
11 New Fetter Lane, London EC4P 4EE

Simultaneously published in the USA and Canada
by Spon Press
29 West 35th Street, New York, NY 10001

Spon Press is an imprint of the Taylor & Francis Group

Typeset in Great Britain by Lucid Digital, Honiton, Devon
Printed and bound in Hong Kong by Tenon and Polert

British Library Cataloguing in Publication Data
A catalogue record for this book is available from the British Library

Library of Congress Cataloging in Publication Data
Tilly, G. P., 1937–
   Conservation of bridges/Graham Tilly; in association with Alan Frost,
Jon Wallsgrove.
      p. cm.
   Includes bibliographical references and index.
   1. Historic bridges–Conservation and restoration–Great Britain. I. Frost,
Alan. II. Wallsgrove, Jon. III. Title.
TG315.T55 2001
624′.2′0288—dc21   2001042027

Contents

Contributors

Graham Tilly PhD, BSc (Eng), ACGI, FICE, CEng, Emeritus Director of Gifford and Partners, Visiting Professor to University of Exeter, has worked in bridge engineering for 30 years. He has written some 80 technical papers and chaired numerous national and international committees. He has special expertise in materials science and carried out considerable research in this area.

Jon Wallsgrove Dip Arch, RIBA was the Highways Agency's Architect and author of 'The Appearance of Bridges and Other Highways Structures' as well as many papers and articles on historic bridges, bridge aesthetics, and, prior to working for HA, he designed major civic buildings and spent 12 years in building construction. He is now Chief Architect of the Court Service.

Alan Frost LVO, AA Dipl, RIBA, DCHM, MaPS has been a member of Donald Insall Associates since 1964, following a Diploma course on the Conservation of Historical Buildings and is now Deputy Chairman of the company. He has particular experience in masonry conservation and in long-term conservation projects at Woburn Abbey and Kedleston Estates, both involving historic bridge strengthening and restoration, as well as having been Architect-in-charge of the post-fire restoration of Windsor Castle.

Foreword

Our national bridge stock is an important asset. Maintenance and refurbishment of old bridges to meet current traffic needs and changes of use is often required. Many of these bridges are important to our cultural heritage and particular skill and care is needed to be able to retain them within a modern transport network and ensure that the solutions developed are sympathetic to their form and appearance.

This book is the first publication to be devoted solely to the conservation of all bridges. The Highways Agency is pleased to have worked in partnership with Gifford to produce it. Gifford is an experienced engineering consultancy in the conservation and restoration of ancient bridges and buildings, and the partnership has brought together the conservation knowledge and engineering skills of Gifford with the research, highway management and practical experience of the Highways Agency.

I commend this publication to all engineers and others who have an interest in roads and the environment. I hope that through this best practice advice those involved in the maintenance and management of all types of bridges will be encouraged to make best use of Britain's bridge heritage.

Tim Matthews
Chief Executive Highways Agency

Preface

This book has been written by Gifford and Partners, Consulting Engineers, in association with Donald Insall, Architects, under a contract to the Highways Agency, London. The terms of the contract were to research and present the best methods of conservation of historic bridges with a view to providing much needed information and advice in an area of growing importance. Information has been collected from a variety of sources including: published case histories, and archives, discussions with practicing engineers and specialist contractors, and the personal experiences of the authors obtained over many years.

The text has been seen and discussed by numerous organisations having interests and responsibilities in historic structures. These include English Heritage, British Waterways Board, The Institution of Civil Engineers' Board of Historic Engineering Works, The Society for the Protection of Ancient Buildings, staff of The Highways Agency, Keith Withey, formerly of Cornwall County Council and John Fisher, formerly Chief Bridge Engineer of Shropshire County Council. More specific advice received from individuals and firms is acknowledged in the relevant chapters. The quality and clarity of the text has greatly benefited from the comments received from these organisations and people. It should be added that in the view of the Society for the Protection of Ancient Buildings this book will be particularly important to County Engineers as they are responsible for the majority of historic bridges in Britain.

The principal author and editor was Dr Graham Tilly. Chapter 9 on Iron and Steel was by Rod Pirie, and Chapter 14 on Archaeology by Gerry Wait, major contributions to chapters 8 and 11 were made by John Simkins and Jonathan Bayliss, all of Gifford and Partners. The Chapter on Architecture was by Alan Frost of Donald Insall Associates.

Many of the photographic illustrations were provided by Dr Graham Tilly, some from his collection but most were specially taken for the work.

Others were taken by Ian Richards also of Gifford and Partners. Photographs from other sources are acknowledged in the text.

The Gifford Project Manager was Tim Holmes supported by Jeanette Hunter, Edmund Hollinghurst, Joint Managing Director of Gifford and Partners exerted a strong influence on the initiation and subsequent strategy of the project. The project was organized and placed by Jon Wallsgrove, Principal Architect, The Highways Agency.

ACKNOWLEDGEMENTS

Figure 3.1 is courtesy of the Tate Picture Gallery. Figs 6.23, 6.24, 6.25 and 6.26 are courtesy Messrs Metalock. Mr Crack (Cambridgeshire County Council) kindly provided information about the Magdalene Bridge, Cambridge. Figs 8.3, 8.10, 8.11, 8.13, 8.15, 8.17, and 8.19 appear courtesy of The Morton Partnership Ltd. Fig. 8.10 is courtesy of the National Trust. Fig. 8.18 is courtesy of New Civil Engineer. Advice provided by the following is gratefully acknowledged: Joe Aley, Cambridgeshire County Council; Edward Morton, The Morton Partnership Ltd, Bethnal Green, London; Mark Sharratt, Assistant Buildings Manager, The National Trust; John Knight, Regional Architect, Historic Scotland. Figures 9.2, 9.8 and 9.30 appear courtesy of The Institution of Civil Engineers. Figures 9.12, 9.16, 9.17, 9.18, 9.19 and 9.33 appear courtesy of Dorothea Restorations Ltd. Figures 9.13, 9.14, 9.18 and 9.36 appear courtesy of Metalock Industrial Services Ltd. Figures 9.26, 9.27 and 9.28 appear courtesy Topbond Group. Chapter 10 – Comments by Professor Roland Paxton are gratefully acknowledged. Figs 12.7 and 12.12 are courtesy Messrs Balvac. Helpful comments by Keith Withey (Cornwall County Council) and Michael Chrimes (Institution of Civil Engineers) are gratefully acknowledged. John Collard (West Sussex County Council) kindly provided information about the Shoreham Harbour Footbridge.

Chapter 1

Introduction

This book has been written primarily for bridge owners and practising engineers, but will also be of value to architects, heritage authorities, archaeologists, specialist contractors and members of the international bridge community. Its aim is to provide information to aid the management and conservation of old bridges within the modern transport network, to improve the quality of conservation and show that demolition can often be avoided. The definition of an old bridge has been taken as one constructed before the motorway era in Britain, that is before about 1960. The advice is derived from examples of conservation work carried out in the past and includes both good and bad practice. The examples are taken from Britain and other countries and deal with relevant problems.

Tasks that have to be addressed in the course of the management of old bridges include routine maintenance, conservation to the requirements of heritage authorities, strengthening to carry higher vehicle loads, widening to meet the higher volumes of traffic and change of use.

Terms such as conservation, rehabilitation, restoration and preservation, tend to be used rather loosely and to avoid misunderstandings it has been necessary to draw distinctions and, for the purposes of this book, define them according to the most generally accepted usage. Summarised versions of the definitions are given in the Glossary of Terms and more comprehensive versions are given here.

'Conservation' is an approach where there is something of fundamental historic or aesthetic merit to be kept, but there can be change, as long as new insertions are in keeping with or enhance that which is existing. It is a living and developing situation. For instance, in a conservation area of a town there could be new buildings as long as they are in keeping with the rest of the environment. Saddling the arch of an old stone bridge or strengthening an existing parapet would be conservation, as would adding contemporary lighting in sympathy with the original design. Re-using an old bridge for pedestrians where it was inadequate for motor vehicles would be a good example of conservation. Within an overall conservation exercise on a bridge there might well be restoration or preservation of certain elements.

'Restoration' is where something has been damaged, changed or mutilated and it is desired to return it to the condition and appearance it formerly had. This needs to be based on good historical evidence, and should use the correct materials. Reinstating a lost parapet or replacing crumbling stonework details would be restoration, as would removing conspicuous and undesirable exposed service pipes, likewise removing an ugly concrete widening to a medieval stone bridge and resiting a separate new structure.

'Preservation' is where everything is preserved in time exactly as is, with no regard for the need to use the item. Corroded and damaged elements would be stabilised and kept as the original material, and not restored. It is the museum approach and is usually considered to be the most extreme. It would be used perhaps for the remains of a Roman bridge. In more ordinary bridges, it would be used for certain elements which no longer have a function, but are of historical interest. Historic signs (e.g. toll charges), damage sustained during historic battles, ancient flood height marks, original tool marking on stones and the ruins of former chapels would all be items to preserve on an old bridge.

'Rehabilitation' is where a structure is of little intrinsic merit, but is worth keeping for sustainability, convenience or economic reasons; there is no imperative to retain the existing appearance. For instance a railway viaduct might be rehabilitated to be used as a highway. Visually there could be enormous changes, often for the better. Entire structural elements might be replaced, e.g. construction of a new deck, and the bridge might be widened or altered in form, e.g. a masonry arch converted to a steel deck using the original masonry abutment. The bridge might even be moved to a different location, with a different span.

'Maintenance' is where routine activities are carried out to prevent or control processes of deterioration such as corrosion, wear of moving parts, decay of timber and degradation of stonework. Activities that are required include removal of rust and painting, cleaning of drains and gullies, lubrication of mechanisms, and repointing of mortar in masonry joints. Maintenance requires inspections at regular intervals and, where necessary, monitoring. Maintenance work that goes beyond routine activities, for example whole scale replacement of machinery in a movable bridge, is referred to as 'refurbishment'.

In Britain, it is Government policy to 'refurbish rather than replace bridges wherever possible' (Highways Agency Maintenance Strategic Plan). Refurbishment would include both conservation and rehabilitation. There are aspects that would suggest that a conservation approach should be taken. These would be age, beauty or character, consistency with its location, whether it was listed or scheduled, whether it had received any design awards, whether it was in or near a conservation area or other sensitive location, or whether it had any literary or historical associations. Rehabilitation would be considered for a bridge without any of the above factors being notably present, but where the keeping of the bridge would have benefits of cost, sustainability or reduced disturbance to users or local people.

The term 'heritage structure' is used to refer to bridges that have been formally listed as having special merit and subject to planning requirements, as explained in Chapter 2. Examples are mostly taken from bridges in Britain, but most countries now have broadly similar legislation in place. Bridges that are not heritage listed, but are nevertheless of historic importance, are referred to as 'historic bridges'. They may merit listing but have been overlooked, or be seen as having only local significance which does not merit national recognition. The various aspects of assessment of the historic merit of bridges are discussed in Chapter 3.

It is important to understand the background to conservation and to this end there are chapters on legislation, attitudes to conservation, architecture and archaeology.

The different types of bridges are categorised according to material and structural form, i.e. masonry, timber, iron and steel, suspended, movable and concrete. These bridge chapters follow a generally similar format, but vary to suit the individual characteristics of the different types.

- A background section provides information about the historical development of the bridge type. This can be useful to aid the various activities required in conservation, for example:
  - identify the age of the bridge and any subsequent modifications or additions

◦   identify the design and mode of construction to aid assessment of the strength of the bridge. This can be particularly helpful in situations when records and drawings are either inadequate or missing altogether

◦   contribute to an understanding of past maintenance and whether it has been sufficiently successful to keep the bridge in good order and operational

◦   identify the historic value of the bridge through its originality, rarity and quality.

The background also provides information about factors such as the impact of utilities on the type of bridge in question and the long-term performance of the bridge type is briefly reviewed.

- The main sections of the bridge chapters are concerned with the methods of conservation. Here the definition of conservation is interpreted to include activities that might, arguably, be considered to be maintenance. For example, painting is generally classified as routine maintenance, but it can also involve some important conservation issues that require to be considered and dealt with correctly. The information contained in the conservation sections provides the engineer with the pros and cons of the various available techniques, and examples of their use, to enable the best solution to be adopted for conservation of the bridge in question.
- Widening is relevant to all types of bridge, but is very common for bridges originally built for horse-drawn vehicles. There is therefore a comprehensive section on widening of masonry arch bridges.
- Strengthening is relevant to all types of bridge, including those built in the 1960s, and is addressed in the respective chapters.
- Summaries of the management activities are given at the ends of the bridge chapters. These sections deal briefly with inspection, common types of deterioration, maintenance and related activities. To an extent this section summarises some of the points made earlier in the chapter.

Where a specific bridge is discussed, the date of construction is given in order to indicate its age. When it can be identified, it is the year the bridge was completed and opened to traffic, even though this has not always been made clear in the references and records consulted. Some writers have preferred to cite the date when the design was prepared or construction was started.

There are a number of features introduced to assist readers wishing to have an overview before reading more deeply into any of the various subjects.

- There are numerous photographs and drawings to illustrate bridges, types of deterioration and conservation work. Many of these illustrations have explanatory notes, some taken from the text and some in addition to the text, to aid readers wanting to obtain a rapid overview.
- There are tables summarising sections of the text to provide abstracts that can quickly be assimilated. At the ends of the bridge chapters there

are tables giving examples of the bridge type, salient details and brief notes on their performance (NB the term 'performance' is used to refer to the success, or otherwise of the bridge in coping with the processes of degradation, the effects of long-term weathering, wind, traffic and pedestrian loading).

- There are bibliographies providing references to the key sources of information. These will assist the reader wishing to go into greater technical depth for specific issues. In preparing this book, the authors have consulted many other sources of information, too numerous to list comprehensively.
- There are glossaries of terms to avoid any misunderstandings in relation to some of the technical issues. The most commonly used definitions have been adopted where possible, but there are also some less well-established terms.

Good practice is relatively easy to identify. Bad practice is less common and is usually due to shortfalls in the general state-of-art in the past rather than ill-considered decisions taken in the case in question. An example of this is the use of cement mortar in place of lime mortar for repointing masonry joints. This was fairly common practice until recently and was in the mistaken belief that it would be advantageous to strengthen the joints. Nowadays cement mortar is being raked out of joints so that lime mortar, similar to that used originally, can be put in its place. The significance of lime mortar is explained in Chapter 7.

Conservation is a 'green' issue, which is concerned with retention and development of the cultural heritage, and with sustainability in construction. These are objectives being pursued by the governments of most countries and are likely to become increasingly important.

Chapter 2

Legislation

Most countries recognise the value of their historic bridges and have legislation to ensure that they are properly managed. In this chapter, the historic development and current legislation in Britain is outlined.

Funding of the maintenance and repair of bridges has always been a matter of concern and dispute. In 1215, Chapter 23 of Magna Carta stated that:

> No manor or man shall be compelled to make bridges over the rivers, except those who ought to do it of old were and rightfully to do so.

This is presumed to cover responsibility for maintenance. In medieval times, there were many inquisitions to decide who should carry out the work and a common verdict was that no one was responsible. In consequence, Grants of Pontage or licences to collect tolls for a limited period, were frequently issued to people appointed to carry out the maintenance. The Statute of Winchester in 1285 affirmed that Lords of the Manor were responsible for the upkeep of the King's Highway. The Statute of Bridges, in 1531, declared that the burden of bridge maintenance should fall on the county. In 1888, the Local Government Act was a major step forward as it made county councils responsible for maintaining all main roads in the country. In 1894, the maintenance and repair of other roads in rural areas became the responsibility of rural district councils. There was, however, no administrative machinery for co-ordinating policies and activities of the councils. In response to pressures from motoring organisations for a central road authority, the Road Board was formed in 1910. This was superseded in 1919 by the creation of the Ministry of Transport (see Jeffreys, 1949). One of the first tasks of the new Ministry was to classify roads to aid decisions on the allocation of finance for their maintenance and improvement. The classifications were:

- Class I, the main arteries and trunk roads
- Class II, other roads of less importance

STATUTORY PROTECTION FOR HISTORIC BRIDGES

All historic bridges possess intrinsic interest and also contribute significantly to the historic character of the landscape. Only some of these bridges are protected under the current legislation. Bridges may be protected by being listed as being of:

- special architectural interest;
- historical interest; or
- scheduled as of national archaeological importance.

Bridges within conservation areas are protected by conservation area legislation, even though they may be neither scheduled nor listed. Others may not be protected at all, but may still possess historic character and interest that is worth preserving. Their conservation needs may differ slightly, depending upon why they are protected, but they should always be conserved in a manner appropriate to their appearance and method of construction.

Bridges were first given legal protection in 1919 as an urgent response to the Ministry of Transport Act which had set aside money for bridge 'improvements'. In the absence of any listed building controls, the Ancient Monuments Board was asked to select 'nationally important examples of bridges constructed prior to 1800' for addition to the Schedule of Ancient Monuments.

The Ancient Monuments Board was created by the Act for the Better Protection of Ancient Monuments, with extended scope in the Ancient Monuments Protection Act 1900, to include 'any structure ... of historic or architectural interest or remains thereof'. Once included on the 'schedule' monuments could be subject to a preservation order and thereby protected from destruction.

The town and country planning legislation of 1947 introduced the requirement to compile lists of buildings (structures) considered to be of special architectural or historic interest. Thus, bridges were protected under two separate legislative codes. These are now enacted under the Ancient

Monuments and Archaeological Areas Act 1979 and the Town and Country Planning Act 1971. Some structures may be both scheduled and listed but, for statutory control, the requirements of scheduling take precedence over listing.

The Town and Country Planning Act of 1971 makes separate provisions for conservation areas, controlling the demolition of historic structures, and requiring local authorities to prepare schemes for the preservation and enhancement of such designated areas. Bridges may clearly be subject to this control, just like other historic structures.

The first lists of buildings of special architectural or historic interest were produced in 1947. The criteria were less restrictive than those covering scheduled ancient monuments, and a wide range of historic bridges was eventually included. In general, bridges that are scheduled for their archaeological interest are often those which are no longer in regular use as part of a highway, such as the Iron Bridge in Shropshire. Despite this, the same bridge may be both scheduled under the 1979 Ancient Monuments and Archaeological Areas Act and listed under the 1990 Planning (Listed Buildings and Conservation Areas) Act. However, in these cases, the scheduled monument legislation takes precedence over the listed building controls.

WORLD HERITAGE STATUS

World Heritage Site status is conferred by UNESCO based upon specialist advice by ICOMOS (International Council on Monuments and Sites), for areas or sites considered to be of outstanding value to humanity. This is embodied in an international treaty called the Convention Concerning the Protection of the World Cultural and National Heritage, adopted by UNESCO in 1972. In Britain, national legislation, as summarised here, is considered to give more than adequate protection. Thus World Heritage site status is in practice honorific, but can be helpful in securing grants for conservation work. Britain currently has some 17 World Heritage sites, of which the City of Bath, Edinburgh and Ironbridge Gorge encompass numerous fine bridges.

SCHEDULED MONUMENTS

Decisions on scheduling of monuments in England are made by the Secretary of State for the Department of Culture, Media and Sport (DCMS), acting upon advice provided by English Heritage (formerly the Historic Buildings and Monuments Commission of England). In Wales, Cadw-Welsh Heritage provides advice to the Welsh Office, and in Scotland, Historic Scotland advises the Scottish Executive. In Northern Ireland, the Environment and Heritage Service fulfils this advisory role. In 1986, English Heritage embarked upon the continuing Monuments Protection Programme (MPP), a systematic and rigorous review of all scheduled monuments. Monuments had previously been selected on an intuitive basis, but MPP is changing the process and ensuring that scheduled monuments are more representative of the nation's surviving heritage monuments and sites. A practical consequence of MPP is a predicted five-fold increase in the number of scheduled monuments.

The 1979 Act defines unauthorised work or damage to scheduled monuments as an offence. It further stipulates that 'scheduled monument consent' (SMC) must be obtained for any works resulting in demolition or damage to a scheduled monument; works which would remove or repair part of a monument; and any flooding or tipping actions on land in, or under, which there is a monument. Applications for SMC are determined by the Secretary of State for DCMS in accordance with the National Heritage Act 1983 and with advice from English Heritage. Exceptional cases will be 'called in' by the Secretary of State to a public inquiry before a decisions is finalised. In addition, the 1983 Act authorises certain works under 'class consents', among which are works:

- executed by the British Waterways Board in relation to land owned or occupied by them, including works or repair or maintenance not involving material alterations, or works which are essential for ensuring the functioning of a canal (class III);
- or repair or maintenance of machinery not involving material alteration to a monument (class IV); and
- which are essential for the purposes of health or safety (class V).

LISTED BUILDINGS

The Secretary of State maintains lists of historic structures according to the Town and Country Planning Act 1971. These lists are intended for the guidance of local planning authorities because, unlike scheduled monuments, listed buildings remain the responsibility of such authorities. The Act stipulates that works affecting buildings listed as grade I and II* are referred to English Heritage for consultation. Buildings are graded I, II* and II (a grade III previously existed but is no longer statutory, see below). All works involving demolition, alteration or extension of listed buildings require a specific grant of Listed Building Consent. Circular 8/87 (appendix IV, subsection IX) provides explicit guidance on the suitability of alterations to listed bridges. It is recognised that there is a difficulty in reconciling the conflicting needs of conservation and current usage:

> Bridges may have importance for their industrial archaeological interest as well as for their architectural qualities. Original fittings, such as lamp posts, should be retained. Functional services, such as sewers, cables and lighting, should be carefully designed and sited.

In general, bridges, which are still in use, are listed rather than scheduled. Procedures for consultation with specified local amenity groups apply to all applications to demolish a listed building. The Secretary of State may also 'call-in' applications for certain classes of demolition for determination. Unlike scheduling, Crown property can be listed and departments must consult the appropriate local authority about proposals to alter, demolish or extend a listed building. Local authorities who own a listed bridge and wish to alter, demolish or extend, must make their applications to do so to the Secretary of State.

A review of the overlap between scheduling and listing of historic bridges has recently been initiated. The initial selection of bridges for scheduling was made prior to the introduction of listed status, and listing alone may be a more appropriate status for some bridges. The selection of bridges for scheduling depends upon assessment of the archaeological potential for such complex structures, the special treatment afforded by scheduling is appropriate for those bridges where successive building phases incorporate evidence for early phases concealed within later builds. In these instances, a continued overlap between scheduling and listing is likely.

One of the most important differences between scheduled and listed status is that listed buildings do not require consent for repairs that do not affect the 'character' of the monument (which would thus constitute an alteration). English Heritage expect rigorous specification and professional control over the execution of repairs for all historic structures.

In general, the survival of historic bridges is both generally documented and quantifiable (excepting areas not yet covered by list review procedures). In England, there are some 525 bridges scheduled as ancient monuments. However, the former Royal Commission of Historic Monuments of England (RCHME) which maintained the list of structures for the Secretary of State as a public record) had some 6,325 listed structures mentioning 'bridge'.

SELECTION CRITERIA

The criteria (nonstatutory) for assessing importance for scheduling purposes were published in 1983 and are as follows:

- survival/condition (degree which survives above and below ground)
- period (examples of monuments of all types which characterise a period should be protected)
- rarity (examples selected to typify commonplace and rare)
- diversity of features (some monuments contain a diversity of features, others only one attribute)
- group value (links to other contemporary monuments, potential (to contain archaeological and historic evidence).

The selection criteria for listing, established in 1970 by the Listing Committee of the Historic Buildings Council, are based upon date and the value of specific examples within types (those showing technological innovation or virtuosity, those with significant historical associations, or buildings with particular group value). Principles of selection approved by the Secretary of State include:

- all bridges built before 1700 surviving in anything like original condition;
- most bridges circa 1700–1840, though some selection is exercised;
- bridges built between 1840 and 1914 are selected according to quality and character, including principal works of principal engineers;
- only selected bridges of high quality built between 1914 and 1939 are listed; and
- a few of the most outstanding bridges, post-dating 1939, are listed; for

example Rhinefield Bridge in Hampshire built in 1950, and Kingsgate Footbridge in Durham built in 1963.

Listed buildings are graded to show their relative importance:

Grade I   Buildings of exceptional interest, 2% of listed buildings
Grade II*   Particularly important buildings of more than special interest, 4% of listed buildings
Grade II   Buildings of special interest warranting preservation, 94% of listed buildings

Assessing the relative importance of historic bridges will involve consultations with the relevant county or district archaeologist and county or district conservation officer. The County or District Sites and Monuments record will provide the first source for basic historical and descriptive information, followed by the photographs in the National Monuments Record kept by RCHME.

The Highways Agency is responsible for some 339 bridges on trunk roads in England built before 1915 comprising: ten originating before 1700, twenty-one scheduled as national monuments; 136 built between 1700 and 1840, including nine national monuments and six grade 2 listed; 202 built between 1840 and 1914, including three national monuments and two grade 2 listed.

The numbers of scheduled and listed bridges on trunk roads built between 1700 and 1840 are surprisingly low since most should be listed according to the criteria applied to historic structures in general. Moreover half of these bridges have not been modified visually. In contrast, there are some 525 bridges in England scheduled as national monuments; this is a much higher percentage of the population of old bridges and suggests that local authorities have a 'richer mix'.

More detailed information about The Highways Agency's bridges is given in the appendix.

BIBLIOGRAPHY

Jeffreys, R. (1949) *The King's Highway*. London: The Batchworth Press.

Chapter 3

_____

Attitudes to conservation

ABBREVIATIONS	
SPAB	Society for the Protection of Ancient Buildings
PHEW	Panel for Historic Engineering Works
PPG	Planning Policy Guidance

This chapter deals with past and present attitudes to the conservation of old bridges.

PAST ATTITUDES

Bridges form an important part of the cultural heritage, but unlike old ruins and museum pieces continue to have a functional purpose and must be maintained so that they can meet that purpose.

Notable bridges have always enjoyed a degree of special attention and on occasions when it was necessary to replace them they were demolished with reluctance and only after considerable debate, for example in the 1930s there was much controversy about the demolition of Rennie's Waterloo Bridge to make way for the present bridge across the Thames in London. A proposal to demolish Telford's Conway suspension bridge in 1958 led to a world outcry, since when it has been closed to traffic and made over to the care of the National Trust. The importance of Ironbridge in Shropshire was always recognised and it is now maintained as a monument, closed to vehicles and only used by pedestrians. On the other hand, Billingham Branch Bridge, Middlesborough, was not recognised as having any historic importance and it was decided to demolish it when the railway line became redundant. In fact, the bridge is one of the first to be welded and is the most significant early example in Britain. In response to intense local pressure the decision was reversed and the bridge retained. However, the majority of bridges were not so highly regarded in the past and there was a tendency to replace them with state-of-the-art structures regarded as a good investment at the time. In recent years this attitude has become less prevalent.

Lesser bridges, despite their importance to society, have invariably been poorly maintained and records show that responsibility and finance have been a constant source of dispute.

In the past, there has been little sentiment for ageing structures and, given the choice, they would have been replaced by new ones. The 17th century diarist John Evelyn was typical in seeing no benefit in old buildings and foolishness in repairing them. The landscape gardner Sir Ewerdale Price used terms such as sublime, beautiful and picturesque in describing the pleasing appearance of decaying ruins. In the 18th century, repair was associated with improvement, but had a reputation for bad taste.

In the 19th century, attitudes changed and restorationists, like Sir Giles Gilbert Scott and the Cambridge Camden Society, took the approach of recording, publishing and restoring. After one such restoration, it was said that the building had been made clean, tidy and restored to a more perfect type. Artists took an interest and found inspiration in modern structures of the time, for example the Brighton Chain Pier was immortalised in the seascape paintings of Turner and Constable, see Fig. 3.1.

William Morris and John Ruskin founded the influential Society for the Protection of Ancient Buildings (SPAB) in 1877. Their purpose was to oppose the activities of the restoration architects and the Gothic Revivalists. The manifesto for the SPAB includes the following paragraphs:

It is sad to say that in this manner most of the bigger Minsters and a vast number of more humble buildings, both in England and on the Continent, have been dealt with by men of talent often, and worthy of better employment but deaf to the claims of poetry and history in the highest sense of the words.

It is for all these buildings therefore of all times and types that we plead and call upon those who have to deal with them to put Protection in the place of Restoration, to stave off decay by daily care to prop a perilous wall or mend a leaking roof by such means as are obviously meant for support or covering and show no pretence of other art, and otherwise to resist all tampering with either the fabric or ornament of the building as it stands if it has become inconvenient for its present use, to raise another building rather than alter or

Fig. 3.1 Brighton Chain Pier by J.M.W. Turner.
© Tate, London 2000/The Estate of Paul Nash

enlarge the old one; in fine to treat our ancient buildings as monuments of a bygone art created by bygone manners, that modern art cannot meddle without destroying. Thus and thus only shall we escape the reproach of our learning being turned into a snare to us, thus and thus only can we protect our ancient buildings and hand them down instructive and venerable to those that come after us.'

The SPAB principles have now become enshrined in the ICOMOS (International Council of Monuments and Sites) Charter and are espoused by English Heritage.

One of the significant outputs of the SPAB was to commission Jervoise, a civil engineer, to survey the ancient bridges of England. This was done river-by-river and published in four volumes in the early 1930s. The survey was a work of great scholarship and remains a classic reference to the present day.

The bridges surveyed by Jervoise were almost exclusively masonry arches and mostly on minor roads in England. The more recent bridges in England, Scotland and Wales, constructed up to the 1930s were surveyed by the Ministry of Transport for an exhibition held in connection with the Public Works, Roads and Transport Congress held in 1933. A permanent record was published as the book *British Bridges* and this too is a classic reference.

In response to an increasing interest, the Panel for Historic Engineering Works (PHEW) was formed by the Institution of Civil Engineers in 1968. The primary purpose of PHEW is to record and publicise notable works and structures that illustrate the history and development of civil engineering. Needless to say, bridges figure prominently in the work. One of PHEW's main outputs is publication of *Civil Engineering Heritage*, a series of five volumes (to date) describing historic bridges and other engineering structures. PHEW organise annual historic bridge awards for repair, strengthening and conservation projects affecting bridges and aqueducts over 30 years old, in England and Wales. These awards have successfully raised the profile and interest in conservation and publicised excellent projects that would otherwise have gone unnoticed. PHEW are also developing a system for classifying historic engineering works (HEWs) into four grades.

ENGINEERING PRESSURES FOR CONSERVATION

Over the years the engineering strategy for managing the bridge stock has shifted from an enthusiasm for replacing old bridges, towards sustaining them by continued maintenance and conservation. There are various pressures that have led to this change in emphasis.

- 'It has become less easy to make a convincing financial case to justify demolition and reconstruction. In the past, financial cases were based on unfactored costs and it could usually be shown that high first-cost could be offset against reduced maintenance costs in the future. However, this made no allowance for the discounting of costs into the future and when this is done a high first-cost is rarely compensated by lower costs in the future. This is a controversial issue, but there is no doubt that a discount rate should be used and the question is whether it should be as high as 6% (the value recommended by the Treasury at the time of writing) or some lower value that enables a more credible calculation to be carried out.

- Arguments that new construction is more durable have become less tenable. Hitherto concrete was considered as not requiring maintenance, but in practice it has become evident that reinforcing steel is susceptible to corrosion and expensive to repair. In contrast, masonry arch bridges have performed superbly over hundreds of years without requiring significant maintenance. Whilst it is not practical to return to masonry construction to any extent, there is good reason for maintaining existing masonry bridges and resisting the temptation to replace them.

- There is a realisation that many of the early types of bridge are much stronger than previously thought. Techniques such as supplementary load testing and more advanced methods of structural analysis invariably show that there is hidden strength. Moreover, the weaker bridges have long since collapsed, many in the late 1800s and early 1900s when steam traction engines became common.

- Demolition and new construction poses questions of logistics as there is invariably a shortage of working space particularly in towns and cities. Moreover, the noise and dust caused by demolition, and the disruption to traffic, pollute the environment and have become less acceptable, if not unacceptable, to the general public.

- Sustainability of natural resources has also become a major issue. While it is possible to recycle some of the products of demolition, there is inevitably some that has to be dumped. New materials are energy intensive. The delivery of new materials and removal of old require journeys by heavy trucks with the consequential consumption of fuel and added pollution. Sustainability has not yet become as important as some of the other factors, costs and safety, but is becoming increasingly so.

- A large proportion of the early bridges, particularly those over 75 years old, have historic interest and many are heritage listed. The listed ones are protected to various degrees so that only prescribed types of work are permitted. Demolition requires an overwhelming case to be made. Ones that are not listed can still attract interest and support from local historians and pressure groups.

Management of old bridges in the context of a modern transport system involves coping with the effects of high volumes of traffic, heavy vehicles and a hostile environment. Inevitably the bridges require maintenance and in some cases strengthening.

ASSESSMENT OF THE HISTORIC VALUE OF BRIDGES

Notable historic bridges are recognised by heritage authorities and classified as national monuments, grade I or II listed structures, as described in Chapter 2. These are required to be properly maintained. However, there are other bridges not listed, but having historic value and also deserving of being maintained according to the requirements of conservation. These are referred to as historic bridges.

When reaching a decision on such a bridge, it will be helpful to determine whether PHEW has assessed and graded it as a historic engineering work (HEW). A bridge that is neither listed nor recorded as a HEW may still be of historic significance and worthy of special attention. In such a case it will be appropriate to assess its value against emotional, cultural and rarity issues, taking into account local factors.

The emotional value to a local community might rest in the bridge being identified with a past industry which has now disappeared. In fact there are many such bridges which now remain as the only physical reminders of industries that have long since disappeared. For example, bridges on abandoned railway lines, bridges on tracks to demolished factories, and bridges across dried-up canals and streams.

Cultural values that can be assigned to bridges include aesthetic, archaeological, setting and age. In this context, attention is invariably focussed on the larger and better known bridges, whereas there are many little known small bridges having merit. They may for example have decorative features of unusual or special merit, be constructed on much older foundations, be located in an attractive landscape, or be of an age which merits attention in relation to other buildings nearby.

Rarity value of a bridge may be through specific features of its design, construction materials or the engineer responsible for its design. For example, it may be the oldest, or the only, surviving example of a particular type of design, the first example of the use of a new material or technique (Billingham Branch Bridge, mentioned earlier, is the first significant welded bridge but was close to demolition before it was recognised), have been constructed with iron from a local foundry or brick from a local brick works, or have been designed by a notable local engineer.

GUIDANCE

In Britain, the Department of the Environment, Transport and Regions has, from time to time, issued guidance on the management of old bridges. In 1925, Circular No. 224 (Roads) addressed to all highway authorities, the Assistant Secretary to the then Ministry of Transport, drew attention to the desirability of considering appearance, as well as the structural issues of ancient bridges. The points made in relation to matters such as archaeological interest, local materials and strengthening are as relevant today as they were in 1925. Circular No. 224 is reproduced on page 19.

It is, however, pertinent to note that in 1925 bridges regarded as being of historical importance were almost exclusively ancient masonry structures. Now, at the beginning of the 21st century, many other types of bridge and more recent constructions, including concrete bridges, are accorded heritage status.

In 1964, a booklet 'The Appearance of Bridges' was issued by the then Ministry of Transport, containing a chapter on Problems of Historical Bridges. As in Circular No. 224, reference was made to the growth in traffic and consequent need to strengthen bridges. The different ways of providing an additional lane on a narrow bridge are addressed:

- construct a by-pass;
- construct a new footbridge alongside and convert former footway into a traffic lane;
- construct a new highway bridge alongside; or
- construct a cantilevered addition to the deck.

Recommendations are made in relation to the aesthetic factors which should be taken into account. Included in these is the question of bridge approaches when a separate structure is built alongside. Detailed attention should be given to the treatment of any new embankment, the space between the old and new work, and the junctions between old and new parapets, etc., at each end. It should be added that, in practice, it is very difficult to design approaches having a pleasing appearance and there is a need for new ideas as there are very few successful examples.

In 1996, the Highways Agency published 'The Appearance of Bridges and Other Highway Structures'. Numerous illustrated examples are provided of bridge widening and re-use. These illustrate that widening and re-use can be designed and carried out to give an aesthetically pleasing conclusion.

Planning Policy Guidance (PPG) Note 15, 'Planning and the Historic Environment' and PPG Note 16 'Archaeology and Planning' were published by the Department of the Environment in 1994 and 1990, respectively. PPG 16 has been an enormously powerful tool used by local authorities to ensure that archaeological and historic remains receive due consideration in the planning control process. Thus, works requiring planning permissions now routinely incorporate archaeological recording work, whether on below-ground remains or on standing structures. PPG 15 has been less immediately influential, but nonetheless strongly reinforces the effects on conservation area planning controls.

Most local authorities have enacted planning control policies designed to supplement national legislation and to enact the nonstatutory guidance notes PPG 15 and PPG 16. Most of these local statutes operate through the planning control process. That is, planning consents carry conditions requiring archaeological/historical recording, and specifying aspects of the required works when archaeological or historical structures are affected. In most cases, therefore, local supplementary statutes will affect only the same cases as national legislation, e.g. repair works are frequently not considered or regulated. However, local authority measures often define less significant grades of historic sites and structures, e.g. sites of district or county significance. When affected by proposed works, such lesser structures may merit recording and use of specified repair techniques, though to less stringent standards than applicable to more important structures.

Circular letter addressed to all Highway Authorities
Circular No 224 (Roads)

MINISTRY OF TRANSPORT
ROADS DEPARTMENT
7 Whitehall Gardens
London S.W.1.
14th March 1925

Sir

I am directed by the Minister of Transport to draw the attention of local authorities to the following observations upon the subject of bridge design:

There are few features, whether of countryside or town, which attract more notice than the bridges carrying roads over stream and watercourses. Many of them possess historical and archaeological interest. Some illustrate the fitting use of local materials by our forefathers, while others provide pleasing examples of modern methods of construction. Of recent years, the rapid increase of traffic has impelled highway authorities to undertake the strengthening of many ancient bridges and the building of many additional structures, with the aid of substantial contributions from the Road Fund administered by this Department.

So far as the strength of such structures is concerned, your Council will be aware that for some years past certain regulations have been prescribed as a condition of a grant from the Road Fund. But it is possible for a bridge to comply with these regulations and yet fall short of the legitimate expectations of the public in the matter of architectural design and suitability to its surroundings.

Colonel Ashley accordingly wishes to impress upon all local authorities, who are contemplating the alteration of ancient bridges or the erection of new ones, the great importance of securing at the outset reliable expert advice upon the design – not merely from the standpoint of the stability of the structure, but also of its proportions and artistic character. Seeing how a long life may be anticipated for public monuments of this class, it will hardly be questioned that every care should be taken to build bridges, and form their approaches, in a manner which will display the sound judgment of the days in which we live.

With this end in view, the Minister wishes it to be generally known that when receiving applications from local authorities for assistance from the Road Fund, he will require to be satisfied that the foregoing considerations have been taken into account. There is no reason to assume that the observance of these principles will add to the cost of construction, for past experience shows that bridges are more frequently criticised for undue elaboration than for well-proportioned simplicity.

I am, Sir,
Your obedient Servant
H.H. PIGGOT
Assistant Secretary

BIBLIOGRAPHY

Department of the Environment. (1990) *Archaeology and Planning*. Policy Planning Guidance Note 16, HMSO.

Department of the Environment. (1994) *Planning and the Historic Environment*. Policy Planning Guidance Note 15, HMSO.

Highways Agency. (1996) *The Appearance of Bridges and Other Highway Structures*. HMSO.

Jeffreys, R. (1949) *The King's Highway*. London: The Batchworth Press.

Jervoise, E. (1930) *The Ancient Bridges of the South of England*. London: The Architectural Press (one of a series of books on the regions of England).

Ministry of Transport. (1964) *The Appearance of Bridges*. HMSO.

Public Works, Roads and Transport Congress. (1933) *British Bridges*.

Chapter 4

_____

Evolution of structural form

In this chapter, a brief outline is provided of the development of the common types of bridge constructed up to 1960 and in everyday use. More detailed information about bridge types, their development and performances over the years, is given in the chapters on timber, masonry, suspended, iron and steel, concrete and movable bridges. This background information is to assist with:

- judgement on whether a structural form or detail is common or sufficiently rare to merit special attention
- judgement on historic value of a bridge
- information about components and types of construction no longer used.

A summary of the different structural forms and the dates they were introduced is given in Table 4.1. Summarised information about the introduction and use of construction materials is given in Table 4.2. A summary of the main styles of design of masonry arch bridges is given in Table 4.3.

The early trade routes followed the contours of the countryside and the most favourable geological strata by a process of experience and intuition. Bridges were only required where it was otherwise unavoidable. In consequence the earliest structures were river crossings where it was not possible to use a ford; this continued until the industrial revolution starting with the canal era in the eighteenth century.

Some of the earliest surviving bridges are the primitive river crossings composed of flat stone slabs laid across raised stones, commonly called clapper bridges, see Fig. 4.1. Three of the better known structures that have survived are at Post Bridge, Dartmoor, Tarr Steps, Somerset and Linton, Wharfedale. They are said to date from the Bronze Age, but this has been in question as they would almost certainly have had to be reassembled from time to time and it would be fairer to say that they are ancient sites. Never-

Table 4.1.   Summary of structural forms

Structural form	Dates and examples
Clapper	Up to c1800. Said to have been constructed in the Bronze Age, but surviving examples such as Tarr Steps in Somerset are unlikely to be this old. The best known clappers are low level and close to the water. Later ones were higher and a little more sophisticated.
Arch	Up to c1900. A compression structure having great strength and durability, suited to materials such as masonry, cast iron and mass concrete. Notable examples include Grosvenor Bridge, Chester 1832 and Royal Bridge, Berwick 1928.
Catenary suspension	c1800 to present day. The availability of cast iron and wrought iron, and the development of wire cables and chains, enabled suspension bridges to be developed beyond the early primitive forms. Notable examples include Union Bridge 1820, Chelsea Bridge 1937 and Severn Bridge 1966.
Cable-stayed	c1800 to present day. A number of pedestrian bridges were built in the early 1800s and some relatively short-span road bridges. The structural form went out of fashion, but was picked up in Germany for the Rhine crossings in the 1950s and in Britain in the 1960s. Examples in Britain include Crathie 1834, Albert 1873 (originally cable-stayed, but much modified) and George Street, Newport 1964.
Box girder	c1850 to present day. Significant experimentation and advances by Hodgkinson, Fairbairn and Robert Stephenson enabled Conway Bridge and Britannia Bridge to be developed. The next significant developments in the structural form were in the 1960s culminating in the aerodynamic section of the Severn Bridge, 1966.
Plate girder	c1900 to present day. Half-through stiffened plate girders of wrought iron were first introduced mainly by the railway companies. Examples of welded plating include Billingham Branch Bridge, 1934.
Lattice girder	c1860 to 1940. Early examples include the ill-fated first Tay Rail Bridge which collapsed in a storm in 1879. There are several from that era still in use. Steel trellis girder footbridges were constructed up to 1940. Examples include Kew Rail Bridge, London 1868 and Gaol Ferry Footbridge, Bristol, 1935.
Truss girder	c1850 to present day. Developed in a wide variety of forms, one of the most common being Vierendeel. Examples include Melton Rail Viaduct, Devon 1871, the Second Tay Rail Bridge 1885 and Carlton Bridge, Yorkshire, 1927

Table 4.1.   Continued

Structural form	Dates and examples

Early composite
  Jack arch
  Trough deck
  Filler beam
	c1900 to 1940. The various forms of early composite decks suited short spans of 3 to 7 m. Although having an element of composite action they were not designed as composite in the accepted sense.

Late composite	c1940 to present day. Decks having steel beams and *in situ* concrete slabs were introduced in the mid 1930s. Precast prestressed concrete beams were introduced in the 1940s and have become a dominant form of construction for spans of up to 27 m.

theless, bearing in mind the date and sophistication of Stonehenge started in 3000 BC, it would be surprising if clapper bridges were not constructed at that time. Clapper bridges are robust, easy to construct and have an appealing simplicity. It is therefore not surprising that they continued to be built up to the eighteenth century. The later clapper bridges are more sophisticated having relatively large masonry slabs simply supported on built-up piers, see for example Fig. 4.2. There is a surprisingly large number of clapper bridges still in use, mainly in remote parts of the country and on routes no longer commonly used. Clapper bridges are commonly narrow and only suitable for pedestrians, but some carry traffic for example Pencarrow Bridge, see Chapter 7. Unlike masonry arches, the later clapper bridges have not been championed to any extent.

Timber bridges have probably been around for as long as clapper bridges but require maintenance at relatively short intervals of time. They are susceptible to decay and fire, and in consequence few have survived in their original form to any great age. Timber bridges have been constructed in more advanced forms over the years, one of the most notable being William Etheridge's celebrated Walton Bridge in 1750 having a span of 40 m and recorded in the painting by Canaletto hanging in the Dulwich Picture Gallery. Some significant timber structures were built during the railway era, for example Walkham Viaduct, 336 m long and 40 m high, by Brunel. At that time Baltic pine could be obtained having an average life of 30 years, some lasting as long as 60 years. By 1934, all Brunel's timber bridges had been replaced. Timber bridges continue to be used in present times albeit on a lesser scale, notable examples being Barmouth Rail Viaduct, constructed 1866, and the 'mathematical bridge' in Cambridge, originally constructed in 1749, see Chapter 8. There are numerous smaller wooden bridges and wood was a popular material for movable bridges, see for example Fig. 4.3.

Masonry arch bridges and aqueducts were built throughout Europe by Roman engineers and there are some notable surviving examples, such as the Alcantara Bridge across the River Tagus and the great aqueduct in Segovia, Spain. In southern Europe there are numerous bridges of Roman origin in everyday use, but having had much rehabilitation over the years. In Britain there were a number of Roman bridges, but only the foundations have

Table 4.2. Summary of the introduction of materials

Material	Dates and examples
Timber	Prehistoric to modern times, the last significant timber bridges were railway viaducts, c1845, replaced in the 1930s. Barmouth Rail Viaduct, 1866, is a significant survivor.
Masonry clapper	Bronze Age to eighteenth century, there are some well-known surviving examples such as Tarr Steps, Somerset and numerous lesser known structures.
Masonry arch	Constructed up to c1900. The long span Grosvenor Bridge, 1832, 61 m, remains in general use and original appearance.
Cast iron	The oldest surviving cast iron bridge, at Ironbridge, is maintained as a national monument. Cast iron continued to be used as a main structural material up to about 1850.
Wrought iron	Introduced c1840, significant bridges include the high level bridge Newcastle and Britannia Bridge across the Menai Strait.
Steel (bolted, riveted)	The first significant steel bridge, Forth Rail Bridge, was built in 1890. Riveted construction continued to be used until 1960.
Steel (welded)	Welded bridges were first constructed in the 1930s. Billingham Branch Bridge 1934 has survived as a significant example.
Mass concrete	Comparatively few were built between the end of the nineteenth century and the 1930s. The oldest survivor, Axmouth Bridge, Devon, 1877, has been refurbished and is now limited to pedestrian use.
Reinforced concrete	Introduced c1870, many are in continued use albeit prestressed concrete has subsequently become more popular. The oldest survivor, Homersfield Bridge, is now limited to pedestrian use. Royal Bridge, Berwick 1928, is an outstanding example.
Precast concrete	Introduced c1910. Popular with the railway companies for footbridges in the 1920s and 1930s.
Prestressed concrete	Introduced c1940, early prestressed bridges were constructed for the railway companies.
Post-tensioned concrete	Introduced c1950. Rhinefield Bridge in the New Forest, 1949, is an early example of this type of construction and is in continued use.

Table 4.3.   Styles and architectural features of masonry arches

AD 50–1100	Semicircular arches in use since Roman times. They were ubiquitous until about 1100. The voussoirs are often of large proportions. There are no bridges of this era extant in Britain, only culverts and remains of foundations.
1100–1500	Pointed gothic arches. The spandrels are seen to be load-bearing walls expressed separately from the structural arch on earlier bridges. The depth of the voussoirs is not less than one tenth of the clear span. Up to about 1500 the widths of the piers on multi-span bridges are nearly as much as the clear spans.
1180–1450	Main construction period for monastic bridges in Britain often identifiable by their ribbed arches, for example Kirby Lonsdale 1275.
1250–1500	Fortifications and chapels, for example Bradford-on-Avon dating from the thirteenth century.
1345–1900	Italian Renaissance introduced the segmental arch, starting with the Ponte Vecchio in Florence of 1345, with a span:pier ratio of 5:1.
1400–1900	Flat-segmented arches having a low rise-to-span. Massive piers and abutments, the thickness of which is often between a quarter and one fifth of the clear span traditionally resist the outward thrust, for example Grosvenor Bridge, Chester 1832.
1500–1750	Span-to-pier ratio of point and semicircular arches increase up to 4:1. Tudor four-centred arches with a small point at the crown were considered at the time less 'barbarous' than the semicircular and pointed arches, but the crown can look awkward and weak, especially if it opened up a little when the centring was struck. Refuges common over piers and cutwaters.
1570–1850	Following the example of the Ponte S. Trinito in Florence 1567, a cartouche or other feature traditionally masks the actual crown on long span bridges. A rise-to-span ratio of 1:7 can be achieved.
1590–1850	Larger span arches express the voussoirs across the spandrels following the example of Rialto Bridge, Venice 1591.
1600–1800	Classical bridges, for example Greta Bridge near Barnard Castle, 1773.
1650–1800	Packhorse bridges, having carriageways up to 2 m wide.
1660–1850	Turnpike roads constructed.
1700–1900	Eighteenth century bridge building boom following growth in road transport and deterioration of old monastic bridges. Semicircular arches revived. Three-centred arches and ellipses used for larger and civic bridges, having the advantage of giving smooth lines to flow into the piers, but give weak crowns, both structurally and visually, since the voussoirs cannot radiate from the centres without restraint cramps or joggled joints.
1720–1735	General Wade's military bridges built in Scotland, e.g. Aberfeldy 1733.

Table 4.3. Continued

1750–1850	Spandrels sometimes pierced to relieve the load and help storm water flow, both on larger semicircular arches and flatter arches.
1760–1830	Canal era, large number of single-span bridges elliptical arches built mostly in brick and typically 5 to 7 m span.
1750–1850	Large civic bridges built in expanding towns and cities, for example Bewdley in 1795.
1825–1900	Railway era, long multispan viaducts, often in blue engineering brick, built across flood plains and river valleys, e.g. Balcombe Viaduct in 1842.

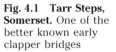
Fig. 4.1 Tarr Steps, Somerset. One of the better known early clapper bridges

survived, the superstructures most probably having been timber which fell into decay after the Romans' departure. There are however a number of surviving subterranean arch structures built by the Romans, one of the most notable being the Monument House Culvert in the City of London, shown in Fig. 4.4, estimated to have had a total length of around 80 m. An excavated length of 20 m was found to be constructed with stone walls having alternate string courses and an arch barrel of tiles fanned out on edge. The section has a span of 0.65 m and height of between 1.30 m and 1.83 m. The culvert is now infilled and preserved.

There are numerous examples of medieval bridges, some built by the monasteries and their 'brothers of the bridge' whose skills are demonstrated by the soaring arches of their monastery buildings. Many of these bridges can be identified by having ribbed arches. Many of the medieval bridges that are still in use, have been widened and strengthened to carry full traffic loading and others have had weight restrictions imposed.

Fig. 4.2 Pophole Mill Bridge, Hampshire. A clapper bridge constructed in the sixteenth century

Fig. 4.3 Wooden movable bridge, Exeter

Bridges built prior to the eighteenth century, can be categorised as those that were wide enough to carry horse-drawn traffic and those that were only wide enough for horses. The latter were the so-called packhorse bridges that were constructed mainly, but not solely, in the period 1650 to 1800 and had a width of no more than about 2 m (6 ft). An example of a packhorse bridge is shown in Fig. 4.5.

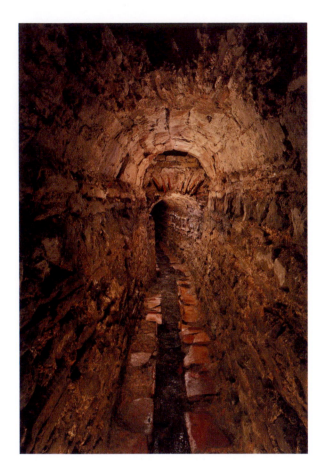

Fig. 4.4 Monument House Culvert, London, c200 AD. A Roman structure in pristine condition after some 1800 years. Courtesy A. Chopping, Museum of London Archaeological Service at MOL

Hinchcliffe identified 189 packhorse bridges in England, many in remote parts of the countryside where they are only used by pedestrians and occasional cattle and horses. Unfortunately, these by-ways attract little or no maintenance funding and the bridges are often in a poor state of repair. This is a real threat to an attractive and valuable segment of the bridge population. Fortunately, some seventy have been listed and must therefore be looked after properly, but there are many others deserving of more care. Many other packhorse bridges have been widened in the past and are now used by modern traffic.

The canal era, around 1760 to 1830, brought a surge in construction of masonry arch bridges, typically of 5 to 7 m span and constructed in brick; stone was also used but less commonly at this time. This was followed by the railway era, around 1825 to 1900, which saw the development of tall multispan viaducts, for example Balcombe Viaduct (1842), across the Ouse Valley in Sussex by Rastrick, and longer spans for example Maidenhead Bridge (1837) by Brunel, having two 49 m spans. On the highways some very fine stone arches were built during the period, for example Coldstream Bridge by Smeaton (1763), Dean Valley Bridge by Telford (1829), and Grosvenor Bridge by Harrison (1832), having a span of 61 m, which remains the longest masonry span in Britain, see Fig. 4.6.

Fig. 4.5   A packhorse bridge. Parapets were normally much lower than in this example

In addition to the technical development of masonry arch bridges, there have been functional and architectural additions. For example, Monmouth Bridge has a fortified gate tower. There are four surviving bridges having chapels, at St Ives, Wakefield, Rotherham and Bradford-on-Avon, the latter was last used as a gaol. Nearby at Bath, there is Pulteney Bridge by Robert Adam (1770), having two-storeyed houses and shops on either side, see

Fig. 4.6   Grosvenor Bridge, 1832. The longest masonry span in Britain

Fig. 4.7. There are a number of bridges having ecclesiastical or other buildings attached which have subsequently been dismantled or partly dismantled leaving remnant foundations, etc.

With the development of landscaped gardens in the eighteenth century, various types of ornamental structure were designed, including bridges across artificial stretches of water. These were invariably covered bridges in the Palladian style as at Wilton, Stowe House and Prior Park, see Fig. 4.8. Others were more functional, but had attractive ornamental features and served to enable gardeners and estate workers to take their equipment across small streams.

Cast iron construction was introduced in the eighteenth century, the oldest surviving cast iron bridge being across the River Severn at Iron Bridge, constructed in 1779 and having a span of 31 m. Tickford Bridge, in Buckinghamshire (1810), is said to be the oldest operational cast iron bridge. It has recently been strengthened using bonded carbon fibre epoxy plating. Other notable examples include Mythe Bridge Tewkesbury by Telford (1823) having a span of 53 m, and there are numerous smaller bridges that have survived, see for example Fig. 4.9.

Wrought iron was introduced in the nineteenth century. One of the first wrought iron bridges was built near Glasgow in 1841, having a span of 9.6 m. The High Level Bridge, Newcastle by Robert Stephenson (1849), had six spans of 38 m. Ribs were of cast iron and tension chords of wrought iron. The most significant wrought iron bridge was the Britannia rail bridge also by Robert Stephenson (1850), having two spans of 70 m and two of 140 m. When designing the Britannia Bridge, Stephenson and his collaborators anticipated the potential problems of steel box girders and an elegant programme of experiments was carried out to confirm the design. The bridge

Fig. 4.7 Pulteney Bridge, Bath. The lesser known upstream elevation

Fig. 4.8   Prior Park, Bath. Palladian-style ornamental bridge. The internal appearance is shown in the inset

Fig. 4.9   Cast iron footbridge across Kennet and Avon Canal, Bath, 1800. A typical example of the elegant ironwork of that era

was the first long-span wrought iron structure and the first significant box girder. It has been said to be the most influential structure of the nineteenth century.

Steel bridge construction was introduced in the second half of the nineteenth century, one of the first being constructed in The Netherlands in 1863. In Britain the Board of Trade imposed a ban which was not lifted until

1877. The Forth Railway Bridge by Fowler and Baker (1890) was constructed in steel plates of 51 N/mm² tensile strength. In contrast wrought iron plates then available were of 34 N/mm² tensile strength. The structural form comprised two 521 m cantilever and suspended spans and it was by far the longest then constructed and remains a gigantic structure for any era. In current times, maintenance of such structures presents new challenges particularly in relation to painting. The traditional red lead paint that has provided excellent protection is no longer permitted due to its toxicity and has to be removed with great care. Likewise the more informal methods of access no longer meet the Health and Safety requirements and more sophisticated methods must be used. The early steel bridges had riveted connections and these were used up to the 1960s, one of the last riveted bridges being Queenhill Viaduct (1965).

In the late 1800s there was concern about the brittleness of cast iron bridges and some were replaced by stiffened plated wrought iron bridges. These were most commonly half-through plate girder bridges on the railways. At the time, steel was available, but the early versions had low fracture toughness and could fracture in a brittle manner. An example of a typical plate girder bridge is shown in Fig. 4.10.

Welding was introduced in the 1930s. The first recorded welded bridge in Britain was across the River Gyfe in Renfrewshire (1931), having a span of 19 m. One of the most significant was the Billingham Branch Bridge (1934), across a railway line at Middlesbrough and having spans of 8.5, 14.6, 19.5, 14.6 and 8.5 m, see Fig. 4.11. In the mid-1980s it became redundant, but local pressure ensured that it was preserved.

The early welded structures experienced a range of problems, including weld cracking during construction, brittle fractures and fatigue. An early

Fig. 4.10 Typical wrought iron plate girder bridge. The stiffeners are attached to the external side of the panels

Fig. 4.11 Billingham Branch Bridge 1934. The first significant welded bridge in Britain

example of brittle fracture occurred in a bridge across the Albert Canal at Hasselt, Belgium. The bridge failed at a weld on a cold day ($-20°C$) in March 1938, just 14 months after construction. Problems due to brittle fracture required considerable research and development before steels having suitably high values of fracture toughness and low brittle–ductile transition temperatures were developed. Fatigue was tackled through recognition of fatigue-prone details and improved quality of welding and weld geometry.

In the 1920s, steel truss girders were found to be more economic than plate girders for long spans. Examples of this era included Carlton Bridge, Yorkshire built in 1927 a viaduct on the A614 having a main span with a through truss of 76 m, and Valley Bridge, Scarborough on the A165 built in 1928 and having four spans of 4.6 m with a supportive lattice truss.

Composite bridge decks having different materials contributing to the structural action were introduced towards the end of the nineteenth century, although they were not designed as having composite action in the modern sense. They are suited to relatively short spans of 3 to 7 m. Many are still in use and present structural analysts with a challenge to demonstrate their true load carrying capacity. Jack arches are composed of longitudinal iron beams having brick arches spanning transversely between the beams and sprung from their bottom flanges or, less commonly, spanning longitudinally from cross-beams. Trough decks are composed of trapezoidal section steel positioned with the stiffer axis in the longitudinal direction or less commonly in the transverse direction. Low strength concrete was usually laid on to the troughing. Filler beam decks are composed of longitudinal steel I-beams encased in concrete. They sometimes had transverse reinforcement and sometimes had nothing. These early types of composite deck were in vogue between 1850 and 1940. The more recent composite systems having reinforced concrete slabs structurally connected to steel beams were introduced in the mid-1930s.

Mass concrete having no steel reinforcement was introduced towards the end of the nineteenth century. The oldest surviving mass concrete bridge in Britain, at Axmouth, Devon, was built in 1877. It is of arch construction having spans of 9.1, 15.2 and 9.1 m. Remedial work was carried out in 1956 and 1989. Despite being located across the River Axe, in a hostile marine environment, the bridge has survived and is in continued use by pedestrians. Glenfinnan Viaduct on the West Highland railway having twenty-one spans of 15 m, was constructed in 1897 and remains in continued use after more than 100 years. At Wansford, an unreinforced arch bridge having three spans of 15, 33 and 15 m, and a total length of 100 m was constructed in 1930 to carry the A1 trunk road across the River Nene. One of the last mass concrete bridges was Pilgrims Way Bridge across the Guildford by-pass in Surrey, built in 1933, but no longer in use.

Reinforced concrete, in its various forms, was also introduced towards the end of the nineteenth century, somewhat later than in other countries. Homersfield Bridge, built across the River Waveney in Suffolk 1870, has a single span of 16.5 m. It has a wrought iron frame encased in concrete and mortar. In 1970, a new road was built and the old road was retained for pedestrians. The bridge was refurbished in 1995 and stands as the oldest surviving concrete bridge in Britain, see Fig. 4.12.

The earliest conventionally reinforced concrete bridge in Britain was built in 1901 to the Hennébique system at Chewton Glen, Hampshire. The Panel for Historic Engineering Works (PHEW) of the Institution of Civil Engineers collected data for 414 concrete bridges built in Britain up to 1914. Of these, the fate of only 187 could be discovered; forty-one were demolished, leaving

Fig. 4.12 Homersfield Bridge, 1870. The older surviving concrete bridge in Britain

Fig. 4.13 Royal Tweed Bridge, Berwick 1928. In the background is Robert Stephenson's Royal Border Rail Bridge

ninety-six extant, of which twenty-six were identified as having been repaired after 1940. Demolition was due to various reasons, such as realignment and low load carrying capacity, as well as deterioration. The fact that there are at least ninety-six surviving concrete bridges that are more than 85 years old is notable. The most common of these early bridges were designed by Mouchel as agent for the Hennébique system and by 1918 some 300 had been built, of which fifty-nine are known to have survived. Reinforced concrete became the dominant form of construction for the next 50 years. Popular structural forms were monolithic beam-and-slab, flat slab, filled arch, open-spandrel arch, portal, cantilever and bowstring. One of the most notable examples of reinforced concrete was the Royal Tweed Bridge at Berwick, 1928, having open arch spans of 51, 75, 86 and 110 m. At the time it was the longest span of its type in the world, see Fig. 4.13.

Precast concrete was introduced around 1910, for example Mizen Head

Fig. 4.14 Precast concrete footbridge, c1928. Although having very low thickness of cover, the concrete is in generally good condition after some 75 years' service

Fig. 4.15   Rhinefield Bridge, New Forest 1950. A grade II listed post-tensioned bridge

Bridge, Cork in Ireland, a through-arch of 52 m span, having precast ribs. Precast footbridges having through-panel girders were constructed by the railway companies in the 1920s and 1930s. A typical example, shown in Fig. 4.14, has weathered rather well and has an attractive appearance after some 70 years' service life.

Prestressed concrete has been around since the early 1900s, but it was not until the late 1930s that prestressed bridge beams were introduced to bridge construction. Two prestressed bridges were constructed for the rail-

Fig. 4.16   Beckton Road Flyover, East London

way companies in 1943, one having I-beams and a span of 13.1 m, the other having box-beams and a span of 15.2 m. Nunn's Bridge, Fishtoft, 1947, having a span of 22.5 m, was the first fully post-tensioned bridge in Britain. Rhinefield Bridge in the New Forest, one of the earliest recorded post-tensioned segmented bridges was constructed in 1950 using the Gifford-Udall system of post-tensioning, Fig. 4.15. It was constructed of lightweight precast units post-tensioned together longitudinally and transversely.

Braidley Road Bridge, Bournemouth, 1968, had external post-tensioning, in which the tendons were positioned outside the concrete section but inside the box structure. Other external tendon bridges built at this time included Cumnor Hill, Oxford and Beckton Road Flyover, see Fig. 4.16.

Movable bridges are an important structural form used on canals, rivers and harbours. There are various types including swing bridges, draw bridges, bascule bridges, rolling lift bridges and transporter bridges. Due to the need for light weight, superstructures are usually constructed in steel for highways or wood for pedestrians.

Suspension and cable-stayed bridges are a structural form used for long spans and for locations where it is desired to have an architectural landmark. There are a number of suspension highway bridges that are 100 to

Fig. 4.17 Union Bridge, Horncliffe 1820. The oldest surviving suspension bridge still open to traffic

Fig. 4.18 Queens Park Bridge, Chester 1923. Typical of suspension footbridges of the 1920s and 1930s, expensive to paint, but otherwise requiring minimal maintenance. Many remain in excellent condition after 70 to 80 years' service

180 years old and continuing to carry traffic, for example Union Bridge, Horncliffe, see Fig. 4.17.

In recent years the early suspended bridges have come to be regarded as heritage structures, requiring special care and attention. There are a comparatively large number of suspension footbridges mostly over 60 years old as very few have been constructed since the 1930s, see Fig. 4.18, for example. Their significance has generally been overlooked and it is timely to take them more seriously. Cable-stayed bridges have become popular in recent years but there are only a few road bridges that are more than 40 years old.

BIBLIOGRAPHY

Hinchcliffe, E. (1994) *Packhorse Bridges of England*. Cicerone Press, Milnthorpe, Cumbria.

Chapter 5

_____

Ancillaries

Items that are ancillary to a bridge are normally nonstructural, but important in an operational sense. With older bridges they may no longer be actively used but have historical or architectural importance, for example notices, chantry chapels, toll-houses, pillars, etc. Or they may be modern additions not provided in the past, but now needed such as access gantries, visitor centres and modern equipment for collection of tolls.

It is important that ancillary items should be retained and kept in a good state of conservation, as they are an important element of the cultural heritage.

NOTICES

The signs or notices attached to bridges, or located nearby, are usually constructed in cast iron and therefore require little maintenance other than occasional cleaning and painting in suitably contrasting heritage colours to enable them to be easily read. The messages given on many signs have become years if not centuries out of date, others are contemporary with the present time. Information on the older signs is invariably of interest as it is informative of bygone days and expressed in quaint English.

Military action and civil unrest have often put bridges at risk and this is sometimes reflected in old notices. For example, during the reign of George IV there were periods of unrest when bridges were damaged sufficient for it to become a transportable offence, as evidenced by Fig. 5.1. There are about sixteen bridges in Dorset still fitted with these cast iron transportation warnings.

In later years, the main risk to bridges was seen to be from the weights of steam traction engines which were many times more than the horse-drawn carts and carriages hitherto. The notice in Fig. 5.2 is typical of those posted at the time.

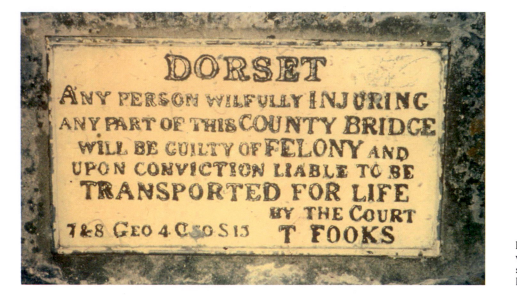

Fig. 5.1 Transportation warning. There are some sixteen bridges in Dorset having this sign

Fig. 5.2 Weight restriction and warning typical of its time. (Warning. Engine Drivers must only take one loaded truck at a time over this bridge, owners and drivers are responsible for damages)

As described in Chapter 10, the early suspended bridges were susceptible to wind-induced vibration and engineers adopted various design schemes in attempts to make their bridges more resistant. James Dredge of Bath was one such inventor and the Victoria Bridge across the Avon was the first of his bridges to be constructed to his patented system, as claimed in the plate shown in Fig. 5.3, see also Chapter 10.

The notice attached to the former toll-booth on Albert Bridge, across the Thames in London, instructs troops to break step when marching across, see Figs 5.14 and 10.2. This is to avoid damage to the deck through the development of high-amplitude oscillations, as discussed in Chapter 10. Although there are many other lively bridges that would qualify for such a warning, this notice is unique.

Whereas the early landmark bridges invariably had the engineer's name displayed in a very prominent manner, later bridges had altogether more modest notices. Typical of these, the suspension footbridges built by David Rowell were simply plated with the contractors' name and date of construction, Fig. 5.4.

Other bridges designed by the professional staff of local authorities simply provide the name of the bridge and the river being crossed, or the names of the various civic dignitaries as in Fig. 5.5, which is typical of the 1920s and 30s.

As the historic importance of bridges has become better understood and public interest has increased, there have been some excellent conservation schemes, some on the initiative of local people or national groups, others by county councils. These have usually been recorded on a suitable plate giving the credits and date of conservation. The plates shown in Fig. 5.6 and 5.7 provide this information in an economical and straightforward manner.

On Homersfield Bridge, Fig. 5.7, there is a coat of arms fixed to the centre of the edge beam, composed of four red hands on a white background. It is believed that it represents bloodied handprints found on a cellar wall

Fig. 5.3 Plate fixed to Victoria Bridge, Bath, 1836. The patent refers to the method of suspension

Fig. 5.4   Gaol Ferry
Footbridge, Bristol

Fig. 5.5   Concrete
bridge, Cambridge

following the murder of a servant in a nearby house in the mid-nineteenth century.

Notices are not always factually correct as at one time no less than six bridges in England had notices claiming to be the first reinforced concrete bridge and dating from 1903 to 1926! In fact the oldest was built in 1901.

Notices providing information about understrength bridges are both functional and important to road users. They are, of course, located so that they cannot be overlooked but, even so, they are not always obeyed. Abuse of weight restrictions is a recurrent theme and weigh-in-motion detectors have been put on trial with varying success. They were used to trigger flashing signs warning drivers that they are OVERWEIGHT and should TURN BACK.

Fig. 5.6 Woodbridge Old, Guildford

Fig. 5.7 Homersfield Bridge, Norfolk. Macabre coat of arms, four red hands, fixed to centre of bridge

The positioning of notices can present a problem, as it is sometimes necessary to have several providing information important to road users competing with each other for attention and having a cluttered appearance. It is usually appropriate for historic plates and notices to be in less prominent positions more evident to pedestrians than motorists, so they do not divert attention from the signs of more immediate importance. In any case, the style of the historic plate, its colour and the size of lettering should be indicative of its content.

CHANTRY CHAPELS AND FORTIFICATIONS

In the years before the dissolution of the monasteries, chantry chapels were sometimes constructed on bridges on the more important routes. They were built as a shrine for pilgrims and travellers. The bridge priest or chaplain would say masses for the dead in return for alms donated by the traveller. This income was intended to pay for the upkeep of the bridge.

These chantry chapels were granted by Parliament to the King in 1545, when responsibility for maintaining the bridges was taken away from the church. Many chapels were demolished and there are now only four surviving in England and Wales, at St Ives, Bradford-on-Avon, Rotherham and Wakefield. As they are integral to the bridges, it is necessary to keep them in a good state of maintenance compatible with the rest of the bridge. They have had chequered histories and nowadays are put to a variety of uses.

The chapel on St Ives Bridge, shown in Fig. 5.8 was consecrated in 1426 and deconsecrated in 1539. It was then used as a private house, two extra storeys being added in 1736. The extra storeys were removed in 1930 when they had become unsafe and the rest of the structure was repaired. The chapel is now used as a museum.

The former chapel on Town Bridge at Bradford-on-Avon, Fig. 5.9, is believed to have been rebuilt or converted to a gaol after the reformation. It is cantilevered off an upstream cutwater and has a stone roof. Later it was used as a store for gunpowder. Nowadays it is divided into two cells

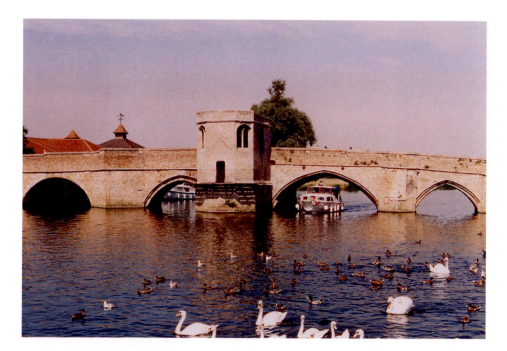

Fig. 5.8 St Ives Bridge. Chantry Chapel now used as a museum

Fig. 5.9   Town Bridge, Bradford-on-Avon. Former gaol that may have replaced a chantry chapel

each containing an iron bed fixed to the walls. It can be visited by special arrangement. The inset to Fig. 5.9 is an informative plate installed by the local preservation trust:

> Town Bridge dates from 13th century and was doubled in width in 1769. The 17th century lock-up may have replaced a medieval chapel. The weather vane is known as The Bradford Gudgeon. The bridge is in the care of Wiltshire County Council.

The chapel on Rotherham Bridge dates from 1483. After the dissolution, it became an almshouse, then a gaol in 1826, then a tobacconist shop, and in 1924 it was restored as a chapel.

The chapel on Wakefield Bridge dates from 1358. It is believed to have been built by King Edward IV in memory of his father, the Duke of York, who was killed in a battle near to the bridge. Nowadays it is consecrated and used for religious services.

The fortified gatehouse on Monnow Bridge, Monmouth, dates from 1296 and was for collecting tolls to raise money for building walls to protect the upper part of the town. It figured in the Civil War and was garrisoned in 1839 at a time of unrest. Llagua Bridge on the Welsh border has what appears to be the remains of a fortification or chapel.

There are records of bridges having fortified gateways, for example Old London Bridge had a gate at its southern end. However, there are no survivors.

TOLL-HOUSES

The funding of maintenance and repair of bridges was often through the collection of tolls. Likewise construction of new bridges was funded in this way. As an aid to the collection, toll-houses or toll-booths were constructed, many of which are still surviving. Unlike chantry chapels, they were usually freestanding and built at the approaches to the bridge. The collection of tolls was gradually phased out, but in 1922 there were still 127 toll bridges. Since then there have been a number of major bridges constructed and funded by tolls, for example Forth Road Bridge and Skye Bridge in Scotland, Cleddau Bridge in Wales, Severn Crossing, Queen Elizabeth II and Humber Bridges in England. There are also historic bridges run by trusts and private individuals that are funded by tolls, for example Clifton Bridge.

Toll-houses were generally built in the style of the bridge, some were plain and others ornate. Some were little more than wooden kiosks, others were permanent stone or brick buildings providing a home for the toll-collector. Where tolls are still being collected maintenance can be funded from the revenue. This is exemplified by Clifton Bridge which, although limited to 4 tonne vehicles, sees a high volume of traffic requiring a modern method of collection. Fig. 5.10 shows the original listed booths and automatic barriers at its western approach.

In contrast, where tolls are no longer collected there can be financial problems as the main concern of maintenance is to ensure that the bridge can safely be used and there is no dedicated income stream to support maintenance of ancillaries. It is desirable for the toll buildings to be used for some purpose, as otherwise they quickly become scruffy, attract vandalism and require additional maintenance.

Toll collection on the masonry arch bridge across the River Avon at Bathampton is by the traditional method operating from the toll-house shown in Fig. 5.11. Current toll charges are displayed prominently. Historic charges displayed on a separate notice contain such gems as: 'Wheelchairs drawn by hand 2d (old pence): Wheelchairs drawn by donkey or pony 3d'.

An example of a former toll-house at Widcome is shown in Fig. 5.12. The original footbridge served by this toll-house collapsed through overloading in 1877. The present wrought iron structure replacing it retained both the toll-house and its nickname of the Half Penny Bridge reflecting the toll charge. The toll-house has three storeys, two being below the level of the bridge deck, and is now a domestic dwelling.

Cleveland Bridge across the River Avon at Bath was constructed with no less than four toll-houses in the style of Doric temples with pedimental porticoes, see Fig. 5.13. They have three storeys with access from the towpath and at the upper road level. Ultimately the income from tolls became insufficient to manage the bridge and as an economy measure only one collection

Fig. 5.10 Clifton Bridge.
Original toll-booths in the foreground and automatic barriers in the background

point was used. The bridge was taken over by the city council in 1925 and tolls were removed in 1929. The toll-houses were disused for many years before being refurbished and modernised as domestic dwellings in 2000.

The wooden toll-booths or kiosks at each end of Ordish's Albert Bridge, in London are in the ornate style of the bridge, see Fig. 5.14. They have been disused for many years and now present a rather forlorn appearance of no longer having a purpose. The painted wooden structures require more maintenance than masonry and inevitably become untidy towards the end of maintenance periods.

VISITOR CENTRES

Visitor centres are becoming an important feature on some of the more significant bridges as they are very popular with the general public and tourists. They provide information about the construction of the bridge, its history and importance as a structure in a format that is easily assimilated by all-comers. Such exhibitions educate people about the heritage and engineering of the bridges. If efficiently managed, visitor centres can cover their cost through entry fees and sale of books, souvenirs, etc.

Fig. 5.11   Bathampton Bridge. Toll collection by traditional means. A list of historic toll charges is also displayed (see inset)

The location and architecture of visitor centres has to be given careful consideration. The best solution is when an existing building contemporary with the bridge can be used in part or in toto. The Clifton Bridge visitor centre is located in a room within a terrace of Victorian houses nearby. Newport Transporter Bridge, Gwent, has a custom-designed visitor centre having an appearance in keeping with the Edwardian architecture of the bridge, see Fig. 5.15. Cheaper buildings in the modern idiom of flat roofs and trimmings of PVC or painted woodwork are not only incompatible and unattractive but have weathered badly.

PILLARS

Pillars or columns constructed at each end of a bridge provide attractive architectural features marking a visual termination of the parapets. Pillars have traditionally acted as anchors to strengthen the run of parapets and protect them from impact from horse-drawn vehicles. They have also formed the gateways sometimes installed on medieval bridges forming part of the city defences. In more recent times, pillars have been erected to support security gates and some of the bridges to Thames islands still have such gates.

For some years, pillars continued to be constructed for architectural reasons to express the solidity and stability of a major bridge, for example Sydney Harbour Bridge has enormous but hollow obelisks for this purpose.

The bridge at Amarante across River Tamaga in Portugal has ornate pillars

Fig. 5.12 Half Penny Bridge, Widcombe. The toll-house is now occupied and still has its original window for collection of tolls

Fig. 5.13 Cleveland Bridge, Bath. Toll-houses now converted to three-storey domestic dwellings

having the intriguing feature that each sits on five stone spheres of 300 mm diameter. The pillars have been used as lamp standards to support lights as shown in Fig. 5.16. Amarante Bridge was the location of hard fought battles against French troops in 1809.

Elvetham Hall Bridge located in a private park in Hampshire, has brick pillars supporting wrought iron gates. The pillars are decorated with stone work and three courses of black bricks contrasting with the light red bricks

Fig. 5.14   Albert Bridge, London. Wooden toll-booth no longer having a public use

Fig. 5.15   Transporter Bridge, Newport, Gwent. Custom designed visitor centre, built 2000

Fig. 5.16   Amarante Bridge, Portugal. Ornate pillars each sitting on five stone spheres of 300 mm diameter

of the bulk of the pillars. They are topped with ornate stone cappings and carved finials, see Fig. 5.17.

The Victorian pillars and remnant foundations shown in Fig. 5.18 mark the location of a three-span masonry arch bridge long since demolished. It is

Fig. 5.17   Elvetham Hall Bridge. Gateway with ornate brick and stone pillars

Fig. 5.18 Demolished Victorian bridge. Pillars and remains of piers marking its location

evident from the angle of inclination of the pillars on the far bank that its foundations were inadequate and this may have been a contributory cause for the need to demolish the bridge. It was a nice touch to leave the pillars, as well as the foundations of the two piers in the water to create a 'ruin' in the appropriate spatial environment of parkland.

Lambeth Bridge across the River Thames in London has tall but rather minimalist pillars with finials typical of the uncluttered design and architecture of the time see Fig. 5.19. This was one of the last bridges to have pillars as they went out of fashion during the 1930s when there was a desire to express the dynamism and speed of motor vehicles. Obelisks and heavy abutments were discarded and designs set out to express the idea of the road streaking off into the landscape. Maillart's celebrated Salginatobel Bridge of 1929 was one of the first to achieve this and led the new movement.

Pillars have both historic and architectural interest and merit preservation on occasions when a bridge has to be demolished. It is not unduly challenging to incorporate them in new structures as they can be free-standing and need not be incompatible with a modern replacement bridge.

PAVILIONS

The architectural desire to express the importance of a bridge and give an otherwise utilitarian structure some presence was sometimes achieved by the construction of ornate pavilions; examples include the Balcombe Rail Viaduct in Sussex and Rochester Bridge in Kent, shown in Fig. 5.20.

Fig. 5.19 Lambeth Bridge, London, 1932. Pillars reflecting architectural details that went out of fashion in the late 1930s

ACCESS FEATURES

Bridges are rarely constructed with built-in access for maintenance work. For the small-span masonry arch bridges, access is usually relatively easy but for longer spans and suspended bridges it can present a problem. Use of specialised mobile access equipment can be helpful in many cases but sometimes the only access is by erection of scaffolding which is expensive and can be disruptive to traffic.

Access is a problem for suspended bridges and particularly the chains and cables. On Clifton Bridge this has been tackled by unobtrusively fitting a small diameter cable to the chains to provide a practical fitment for a safety harness, see Fig. 5.21. Clifton Bridge also has a mobile gantry to enable the underside of the superstructure to be inspected and maintained. This is a luxury rarely enjoyed by historic bridges.

On the original construction, access to towers was usually by ladders having no safety cages and exposed to the wind and rain. The Transporter Bridge at Newport Gwent was accessed in this way and it was just possible for inexperienced climbers, such as the author, to inspect the upper levels of the structure. In recent years access has been upgraded, the stairs and ladders having been fitted with handrails and mesh panels, and the timber

Fig. 5.20 Rochester Bridge. Ornate pavilions give this bridge a presence that would otherwise be absent

Fig. 5.21 Clifton Bridge, Bristol. Fitment for a safety harness to aid inspection and maintenance

walkway at the top has been replaced by mesh flooring to provide a non-slip surface in all weathers. Such changes have to be designed to have minimal impact on the original structure and on structures such as this must receive the approval of the heritage authority.

The Forth Rail Bridge, famous for its painting cycle, was originally accessed by techniques imported from the sailing clippers, but no longer acceptable to the Health and Safety Executive. It has therefore been necessary to revise the procedures to meet present day standards.

The importance of having access to carry out painting is exemplified by a steel structure near Clapham Junction carrying signalling equipment over a main railway line which partially collapsed some years ago because it was inaccessible and the unpainted parts of the structure had corroded excessively.

Chapter 6

Parapets

GLOSSARY

Balustrade	Ornamental parapet most commonly of cast iron, stone or concrete composed of a horizontal coping member supported by vertical pillars termed, balusters.
Baluster	Vertical members in a balustrade, sometimes referred to as spindles. Balusters usually have a circular cross-section of varying diameter being greatest at around mid-height.
Die stone	Solid members positioned at the ends and, for longer spans, at intervals between balusters in a balustrade. Provide an appearance of strength which is not necessarily achieved.
Pilaster	Similar to a die stone, but thicker than the rest of the parapet. Also a feature of solid parapets.
Railings	Lightweight balustrade having slender members of iron, steel or wood, often prefabricated.
Coping	The top course of stone on a masonry wall or balustrade.
Pedestal	The bottom course of stone in a balustrade
Refuge	Recess formed by widening the parapet and pavement to enable pedestrians to shelter from traffic. In later bridges refuges were more of an ornament than necessity.

BACKGROUND

From local records it is evident that many of the early bridges, clapper bridges and masonry arches, were narrow and constructed with parapets that were little more than a kerb. One of the reasons suggested for having low parapets is the fear that low slung packs on pack horses could hit a higher parapet. Another likely reason is the need for economy particularly in the poorer parts of the country as the funding of bridge construction and maintenance has invariably posed problems. In some places the bridges had parapets from an early time, for example there is documentary evidence of there being parapets on Bedford Bridge in 1526. Over the years, as road systems have been developed to meet increased traffic, bridges have been successively widened and parapets have been added. In some cases the retrospective addition of masonry parapets is evident from the different styles of construction in relation to the original. Nevertheless, there remains a number of bridges having no parapets usually in remote locations, for example Passfield Sluice Bridge, a seventeenth century structure in Hampshire, see Fig. 6.1.

Many of the early parapets were timber as evidenced by the three-span masonry arch bridge at Uckfield in Sussex c1850, as shown in Fig. 6.2. Tilford Bridge in Surrey still has a wooden parapet which was carefully refurbished in 1998, as shown in Fig. 6.3. Wooden parapets were often constructed with outstands to provide lateral stability and located outboard of the structure to leave a maximum useable width of carriageway. At this time parapets would have been erected to provide people with assurance, particularly under windy conditions when there could be a fear of being blown off and, more likely, at dusk or darkness when they could simply walk off in error as lighting did not come until the late nineteenth century, if at all on the country bridges.

Simple wrought iron railings were sometimes fitted retrospectively as on Barton Clapper Bridge in Wiltshire, shown in Fig. 6.4. Here, there was no

Fig. 6.1 Passfield Sluice Bridge, Hampshire. A seventeenth century structure having kerbs at its edges but no parapets

Fig. 6.2 Uckfield Bridge, Sussex. Wooden parapets as sketched in the late 1800s, the bridge was dismantled in 1859 to make way for the railway

Fig. 6.3 Tilford Bridge, Surrey. The timber parapets were refurbished along with the rest of the bridge in 1998

evidence of either timber or masonry parapets having been fitted before-hand. The wrought ironwork is very plain along the length of the bridge, but has simple decorative details at each end. The longitudinal flexibility of the railings has the advantage that minimal thermal stresses develop. Connections between the vertical and horizontal rails are by upsetting of spigots.

Fig. 6.4 Barton Clapper Bridge, Wiltshire. There is no evidence of parapets prior to installation of plain wrought iron railings which are in harmony with the stonework. The insets show details of the ironwork

Whereas the wrought ironwork on Barton Clapper Bridge is typical of economical rural workmanship added retrospectively and in harmony with the stonework and age of the bridge, the concrete posts and steel rails added to a sixteenth century culvert, shown in Fig. 6.5, are clumsy and out of proportion.

The beautifully detailed ironwork on Avington Park footbridge, a private bridge in Hampshire, is a very fine example of parapets designed as part

Fig. 6.5 Parapet railings added retrospectively. The railings are clumsy and out of proportion to the small culvert

of the bridge. The ornate bronze fittings and adornments are particularly attractive as shown in Fig. 6.6.

Stone parapets are very common for masonry arch bridges and have performed well over the years. Most were originally constructed using a lime-based mortar having sufficient flexibility to enable the assemblage to cope with thermal expansion and contraction. The apparent solidity of masonry parapets can be illusory because there is little adhesion between the stone blocks and they can fly apart when struck by an errant vehicle. However, energy is absorbed in the collision which is good for passengers, but bad for the bridge. Stone balustrades have little resistance to impact as few have any structural connection between balusters and the horizontal courses of stone.

Brick parapets were commonly used in the nineteenth century, particularly for canal bridges and rail bridges. As for stone, brick parapets having lime-based mortar could cope with thermal effects. However, more recent construction and maintenance using mortar having Portland cement is relatively inflexible and is likely to develop thermal cracking, as shown in Fig. 6.7. The use of cement mortar is now unacceptable in conservation work.

A few months after the photograph was taken the parapet shown in Fig. 6.7 was destroyed in the early hours of the morning by a hit-and-run driver of a truck.

Ornate cast iron parapets were constructed on the more prestigious bridges where it was possible to have a more expensive and eye-catching effect. These were typically river crossings through cities where there was a question of civic pride and sufficient finance to enable it to be achieved. Some very attractive cast iron parapets were constructed having intricate detailing. However, the solid and robust appearance is deceptive and few would meet modern requirements of containment if struck by errant vehicles. Also, the longer spans did not always have adequate detailing to cope with thermal effects. An example of thermal cracking is shown in Fig. 6.8. In this case the transverse joints between the cast sections have been stepped, presumably for the sake of appearance, but inadequate allowance was made for thermal movements. The hollow cast iron balusters were subsequently repaired; corroded internal reinforcement bars were sprayed

Fig. 6.6 Avington Park footbridge, Hampshire. A fine example of decorative iron parapets

Fig. 6.7 A reconstructed brick parapet using Portland cement mortar. Cracking has occurred due to the lack of flexibility

with protective material unless they had lost a significant percentage of cross-section in which case they were replaced. New castings were fitted in place of the cracked ones.

Fig. 6.9 shows a fine example of a lightweight cast iron parapet where the palings are cast as arrows to produce an attractive and well-balanced appearance. Rather ominously, the notice apparently advises people not to stand on the bridge for reasons of safety, but this actually related to traction engines in earlier years.

Cast iron became cheaper than masonry and was sometimes used in preference for example Chertsey Bridge built across the Thames in 1782 had iron balconies as refuges because they were cheaper than extending the cutwater. However, the iron brackets fractured very quickly and were removed, and now only some of the ornamental panels remain incorporated in the masonry parapet.

Cast iron and wrought iron have the great advantage that they are fairly resistant to corrosion. In contrast, steel is rather susceptible and requires regular maintenance, cleaning and repainting. Riveted lattice steel construction became popular towards the end of the 1800s. Unfortunately corrosion can occur at the interfaces where the inclined members are riveted together, and also on flanges where multiplates are riveted together to provide added bending strength. On occasions when there has been corrosion at the connections of inclined members, rivets have been replaced by bolts. This may have been unnecessary unless there was good evidence that there was corrosion at the interfaces as opposed to the external faces. An example of a seriously corroded flange assemblage is shown in Fig. 6.10. The expansive force developed by corrosion products eventually fractures the rivets causing the structure to be weakened. If the corrosion is not too extreme, the normal treatment is to replace the broken rivets with bolts and tighten them

Fig. 6.8 Cast iron parapet. Cracking has occurred in the baluster due to restraint preventing the parapet from freely expanding and contracting

to a force high enough to compress the interfaces together so that moisture and air are excluded. The steelwork is then painted taking care to seal the joints. In practice, this requires great care, otherwise it will not produce a lasting solution. The flange shown in Fig. 6.10 is in an advanced stage of interfacial corrosion and has reached the point where the plates should be dismantled, cleaned, treated and reassembled. An alternative and less satisfactory treatment is to remove the plates, clean the parent flange and weld a new plate in place, taking care to avoid the creation of fatigue problems that can initiate at the toes of the welds, particularly at transverse welds.

Wrought iron-plated parapets have been used for railway bridges where the stiffened plates act primarily as longitudinal beams supporting the deck. These are easier to maintain than lattices as maintenance work is more straightforward. Positions of the riveted joints are less susceptible to corrosion and durability is generally better. An example of a steel plated parapet is shown in Fig. 6.11. There is very little corrosion to be seen above deck and the steel joints have weathered very well. However, it has been damaged by vehicle impact and several of the vertical web stiffeners have been badly distorted to the extent that the compressive stability has been reduced. Both lattice and plated parapets are susceptible to corrosion at

Fig. 6.9 Wimborne Bridge. The arrows give a nicely balanced appearance, the notice reads 'Safety First. Do Not Stand on Bridge'

Fig. 6.10 Steel lattice parapet. Interfaces of the cover plates on the top flange have reached an advanced stage of corrosion causing expansion and distortion

the interface between the plating and the asphalt surfacing of the roadway or pavement. At this location detritus and de-icing salt can collect during winter maintenance and there are cycles of wetting and drying which provide ideal conditions for corrosion to occur.

Fabricated steel railings introduced in the 1920s have performed well.

Fig. 6.11 Plated wrought iron parapet. Several of the vertical stiffeners have been distorted by impact of an errant vehicle. This has weakened the compressive stability. The inset shows corrosion at the vulnerable interface with the asphalt pavement

They have the advantage that there are no interfaces or other traps where corrosion can occur in steelwork above deck level. They are fairly easy to maintain as they are accessible for cleaning and painting. There are, however, corrosion traps on some types of connection to the deck.

Concrete parapets were introduced in the early 1900s. They followed the normal course of events and copied the structural forms of the established masonry parapets. Some were solid like a masonry wall, some were composed of cast balustrades like the more prestigious stone ones, for example see Fig. 6.12. Others took advantage of the possibilities of reinforced con-

Fig. 6.12 Concrete balustrade. Constructed in 1926 and in reasonable condition after 74 years exposure to weathering and de-icing salts

crete to have structural forms in a more modern idiom, see Fig. 6.13. Later designs incorporated precast units, some being composed of relatively thin panels. The thickness of cover concrete over the steel reinforcement was generally very low and by current understanding provided little protection from corrosion. Nevertheless, performances have been surprisingly good and there are many surviving precast concrete parapets that are 50 to 100 years old. However, few are sufficiently strong to satisfy current requirements for vehicle impact.

Composite parapets having concrete posts and steel rails were also used. The example shown in Fig. 6.14 is of unusually economic construction typical of the 1940's. Steel mesh has been added at a later date to conform to updated requirements. The onset of corrosion in the reinforcement steel in the posts is not surprising bearing in mind the low thickness of cover and exposure to over 50 years weathering and de-icing salts.

Parapets on suspension bridges were sometimes designed to contribute structurally having the hanger cables attached to the top flange of a truss or plated parapet. Examples include Albert Bridge and Chelsea Bridge, London. This system has the advantage that the hanger connections and structural members can more easily be inspected and maintained. Steel suspension footbridges, such as those built in the 1920s and 1930s by David Rowell had lattice trusses having the dual role of stiffening the superstructure and acting as parapets.

Some precast concrete footbridges, also constructed in the 1920s and 1930s, had tall side panels that acted structurally as beams as well as parapets. They invariably had overhead bracing to provide lateral stiffness. The bracing could also be used to carry lighting. The side panels incidentally

Fig. 6.13 Concrete parapet. Constructed in 1926 in a more modern idiom. The concrete has weathered attractively

Fig. 6.14 Composite steel-concrete parapet, 1948. Corrosion has commenced in the steel reinforcement

blocked the view from the footbridge providing the pedestrians with a rather gloomy tunnel-like environment.

REQUIREMENTS OF PARAPETS

Parapets on the older bridges were designed as guards to delineate the edge of the bridge and provide a degree of protection against horse-drawn vehicles breaking through. Design strength was a matter of judgement by the engineer and the local requirements of the bridge owner.

Nowadays, it is recognised that parapets are subject to a variety of types of loading and are required to fulfil a number of requirements, some of them occasional, depending on usage and location. These can be summarised as follows:

- Structural contribution to strength of the bridge
- Containment of vehicle impact
- Containment of crowds of pedestrians
- Wind and its effects
- Protection against criminal acts
- Restraint of horses

- Restraint of suicides
- Good aesthetic appearance.

Many of the parapets constructed during the railway era were designed as part of the load-bearing structure, for example half-through and through-girder bridges. This is an attractive and economic structural form that enables the depth of construction to be minimised, but is vulnerable to vehicular damage. It is acceptable on railway bridges where there is a low likelihood of derailment and impact, but the situation is different on highways. In a fatal accident on a highway bridge in Alabama, USA, a car collided with a petrol tanker causing a conflagration which weakened the steelwork and led to collapse of the bridge. Where it is feasible, standard barriers are retrospectively installed to protect the members.

In Britain, vehicle containment standards were introduced in 1966 and subsequently made mandatory for bridges over railways, over and under motorways, and trunk roads, and were recommended for bridges on other roads.

Existing parapets constructed prior to 1966 that have not been retrospectively strengthened are therefore liable to receive impacts that would not have been taken into consideration when they were designed. The resulting damage can be merely cosmetic, structurally weakening or loss of a section of parapet. This 'parapet bashing' is akin to the better known 'bridge bashing', when high-sided vehicles attempt to pass beneath low bridges causing damage to masonry arches and edge beams.

Some of the masonry bridges on minor roads are particularly susceptible to bashing when there are bends in the road at the approaches to the bridge to enable a right angled crossing to be constructed. In the United States, these are called S-bridges. They present a problem to drivers of long vehicles as special care has to be taken to avoid the tail end from bashing the parapet at an oblique angle as they run on and off the bridge. In extreme cases a section of a masonry parapet can be pushed off. In less extreme cases, the damage can be more cosmetic as illustrated in Fig. 6.15, where the brick parapet has been scraped by passing vehicles on many occasions without apparently causing any serious damage. On the parapet shown in Fig. 6.16, a section has been knocked off altogether and the end pillar has been knocked sideways. Such damage is potentially dangerous to people who may be beneath the bridge and is expensive to repair. Moreover, it invariably recurs at intervals. Parapet bashing is becoming so common and expensive to repair that there is a need to take action to minimise it. However, erection of warning signs and painting black and yellow lines, as for bridge bashing, are not appropriate to historic bridges as they would be ugly and damage the setting. Signs located well away from the bridge giving early warning could be helpful. Parapet bashing is a problem that would benefit from research.

The parapet bashing illustrated in Fig. 6.11 for a steel-plated structure had involved a more serious impact when a vehicle had crossed the road and exerted high forces on the vertical stiffeners of the steel plating.

Crowd loading of parapets can occur when spectators are gathered to watch events taking place near or below the bridge. Horizontal forces can develop perpendicular to the parapet which could dislodge masonry or

Fig. 6.15 Parapet bashing. The brickwork has been scraped on numerous occasions without causing structural damage

Fig. 6.16 Parapet bashing. A section of this parapet has been knocked off and the end pillar deranged in separate incidents. Parapet bashing is becoming an expensive and common problem

break wooden structures, particularly if they have aged and weathered. This is of concern to authorities having bridges which are regularly exposed to popular events, such as sporting activities. In some cases the bridges have to be closed to spectators on the day of the event as a precautionary measure. In others, investigative work is being carried out to develop a method of strengthening; as discussed later in this chapter. Damage can also be caused by people climbing on to parapets and jumping into the river below as a 'recreational activity', or more likely, bravado. This is dangerous to the people concerned and has occurred on numerous bridges and related structures, for example the embankment by Pulteney Bridge, Bath. Notices forbidding such activities are not uncommon.

Parapets interact with wind in two ways: they are a significant part of the edge profile and therefore contribute to the aerodynamic stability of the superstructure, and they affect the impact of the wind on pedestrians. In general, wind is only a significant factor on bridges in exposed locations. The interaction between parapets and wind has rarely been taken into consideration and in any case would require expensive wind tunnel tests to resolve. Nevertheless, it is a factor to be considered if parapets have to be modified for one reason or another.

In recent years there has been a growth of criminal acts on bridges over railways and roads, where people drop heavy things such as concrete blocks on to vehicles passing beneath. This has led to the necessity of constructing higher parapets and enclosures on bridges in locations at risk. Such locations are mainly over motorways. The quieter areas where old bridges are usually located have been relatively free of this problem.

Access bridges over busy roads in rural areas where the fast moving traffic below could cause horses to become disorientated and panicked, sometimes have high equestrian parapets. This mainly involves bridges over motorways and is unlikely to be relevant to old bridges.

High bridges crossing dramatic locations tend to attract people intent on committing suicide by jumping off. This has been recognised as a problem for many years and various efforts have been made to tackle it. Clifton Suspension Bridge, Bristol, is a notable case where the Bridge Trust has erected notices providing information about the Samaritans in an effort to dissuade suicides by counselling. Furthermore, a public competition was organised to invite schemes for modifying the parapets without significantly changing their appearance. The result, shown in Fig. 6.17, was a cleverly conceived design, having horizontal stainless steel wires to provide a physical barrier which made little change to the appearance of the parapet when viewed from the pavement and none from the banks either side of the bridge. On Orwell Bridge in Suffolk there is a telephone system providing a link to the Samaritans.

Telford's masonry arch bridge across Dean Valley in Edinburgh was another that attracted suicides; in this case they were stopped by simply raising the parapets by 0.3 m.

APPEARANCE

Parapets play a significant if not crucial role in the appearance of bridges. Prior to the standardisation, which commenced in the 1960s, they were designed individually to be compatible with the rest of the structure. This has resulted in a rich heritage of parapets reflecting the different eras and uses of materials. Unfortunately parapets have not always been allocated sufficient maintenance funding due to more pressing needs and, on some of the more remote bridges, were allowed to become scruffy and damaged.

Fig. 6.17 Clifton Suspension Bridge, Bristol. Parapet having a suicide barrier fitted retrospectively. In the background is the gantry providing access for inspection and maintenance of the underside of the bridge

In more recent times this has been redressed to some extent and there are notable examples where damaged or lost sections have been replaced to match the originals, corrosion has been treated and ironwork repainted. Examples are given later in this chapter.

In addition to the repetitive details along their lengths, parapets often have ornamentation and features such as the coat-of-arms and name of the local authority, date of construction, names of the engineer and contractor, etc, as exemplified in Figs 6.18 and 6.19.

Parapets sometimes incorporated refuges where pedestrians could shelter from the traffic. In medieval times, on bridges having no pavements, refuges were probably essential to safety, but the more modern versions built in Victorian times are ornamental and nowadays serve to provide people with a space where they can pause to enjoy the view, if they can spare the time. On narrow bridges refuges continue to have functional value as they can obviate the need to widen the structure to provide dedicated lanes for pedestrians, cyclists and horses. New refuges can be created on approaches to existing narrow bridges. Examples of refuges are shown in Figs 6.20 and 6.21.

Sadly, refuges are no longer constructed and bridges are seen as a part of the road to be traversed without pause. Modern parapets are invariably designed with their predominant members horizontal and having an appearance similar to the rest of the road, so that bridges have become anonymous and unnoticed by travellers.

CONSERVATION

Conservation has become an important issue as it has become recognised that it is important to retain the appearance and originality of parapets.

Fig. 6.18 Mights Bridge, Southwold, Suffolk. The feature embodied in an otherwise rather plain parapet has provided added interest and improved the appearance

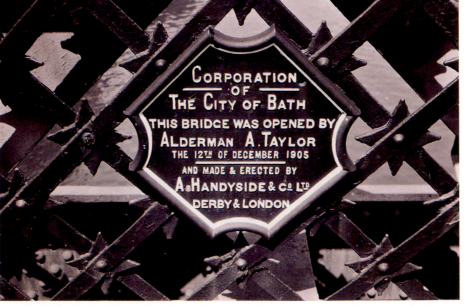

Fig. 6.19 Commemorative notice, Bath. Plate giving typical information and constructed to match the decorative detailing of the lattice parapet

Specialist restoration firms have tackled the problems and some interesting and innovative schemes have been carried out. Some of the more common methods of conservation used in practice are listed in Table 6.1. Clearly, the different structural types and materials require different approaches as outlined in the following sections.

Fig. 6.20   Prebends Bridge, Durham. Example of a traditional refuge providing stunning views of the river valley

Fig. 6.21   Blackfriars Bridge, London. Example of a Victorian refuge, having a red granite column

Timber

In principle wooden parapets can be maintained indefinitely in their as-built form, provided due care is taken to ensure that ageing and deteriorated members are replaced like-for-like and in good time. Deterioration of trusses and parapets invariably initiates at connections and unprotected end grain,

Table 6.1.   Methods of conservation of parapets

Defect	Method of conservation	Advantages	Disadvantages
Deterioration of timber	Replace like-for-like.	Retains appearance.	None identified.
	Cut off deteriorated ends and recycle shortened members.	Retains originality.	
Cracked iron and steel	Weld.	Retains appearance and originality.	Cracking will recur unless underlying causes are tackled.
	Cold stitch.	Retains appearance.	
Corroded iron and steel	Clean, remove rust and contaminants. Re-paint.	A normal maintenance activity.	Corrosion will recur unless underlying causes are tackled.
	Replace weakened members like-for-like.	Retains appearance.	Loss of original material. Expensive to produce cast iron or wrought iron.
	Replace non-structural members by fibreglass replicas.	Retains appearance and economic. Corrosion will not recur.	Unlikely to be acceptable to heritage authorities.
	Cut out corroded part and weld in place new material.	Little change in appearance. Effective and economic.	Corrosion will recur if underlying cause is not tackled.
	Remove rivets and cover plates, clean the interfaces and treat with a suitable protective paint. Reassemble and preferably fit rivets.	Little change in appearance. Economic.	Corrosion will recur unless the interfaces are thoroughly cleaned.

Table 6.1.   Continued

Defect	Method of conservation	Advantages	Disadvantages
Damaged, stolen or vandalised iron members	Replace like-for-like.	Retains appearance.	None identified.
Deteriorated mortar joints in masonry	Repoint with lime mortar.	Traditional method.	None identified.
Deteriorated masonry	Replace like-for-like.	Retains appearance.	None identified.
	Replace coping with concrete of similar dimensions and colour.	Can weather in a compatible manner. Can provide better durability.	Unlikely to be acceptable to heritage authorities.
Corroded concrete reinforcement.	Clean and patch repair.	Minimal disturbance to original materials.	Appearance may be changed and likely to be incompatible with original concrete.
	Clean and coat with sprayed concrete.	A practical method for thin precast sections.	Appearance changed and incompatible with original concrete. Unlikely to be acceptable to heritage authorities.
Bent or distorted steelwork	Straighten by local heating and constraint.	No change to appearance or material.	Requires knowledge, skill and experience.

typically at connections to the deck where moisture can collect. It follows that for longer members having deteriorated ends it is sometimes practical to cut off the defective material and recycle the resulting shortened members in other parts of the structure. In Fig. 6.22 an example of a weakened parapet is shown. Temporary fencing has been erected to prevent people from getting too close.

Fig. 6.22 An example of a deteriorated timber parapet. Palings in the foreground have been erected to keep pedestrians back

Iron

Repairs of cast iron and wrought iron parapets require careful attention to detail in order to avoid losing originality and spoiling the appearance. Cracked or broken cast iron can be repaired by cold stitching or welding, as described in Chapter 9. The cold stitching process, illustrated in Figs 6.23 to 6.26, has been used successfully on a number of occasions to repair

Fig. 6.23 Cast iron parapet before repair. Rear view

Fig. 6.24   Replacement panel. Front view

Fig. 6.25   Cast iron parapet repaired by cold stitching. Prior to final painting

cracked cast iron but is a highly specialised activity requiring a full understanding of the technical processes and experience.

Damaged or missing sections of cast iron parapets can be replaced by using a remnant section as a pattern to enable new moulds and castings to be made. It should be noted however that, when using an existing section

Fig. 6.26 Completed parapet repair. After final painting

as pattern, allowance must be made for the change in size due to thermal effects. As an example, the cast iron parapets on the historic Homersfield Bridge in Suffolk (see Chapter 12) had deteriorated and been vandalised over the years. Frost had weakened some of the footings, York stone copings at the edge of the deck were damaged, and sections of the cast iron parapet were missing. As part of the conservation work the stone copings were removed and cleaned. The cast ironwork was also removed and restored. Where sections were badly damaged or missing, moulds were made from the best of the survivors and new castings produced. The refurbished parapets, shown in Fig. 6.27, are very fine and the repairs are indistinguishable from the originals.

Masonry

Conservation of masonry parapets presents no particular difficulties and is mainly a matter of replacing deteriorated stones like-for-like and repointing with lime-based mortar. As for masonry arches, there have been some rather poorly executed repairs in the past using ill-matched materials. For example, bricks inserted into stonework and different types of brick into brickwork. There have also been occasions when masonry coping stones have been replaced with concrete. Actually, concrete coping stones can turn out to be surprisingly successful as a reasonable colour match can usually be obtained and weathering can be compatible. The differing characteristics of concrete can be tolerated more readily in substituted copings than in the mass of a masonry wall because they are located on a natural boundary and have less interaction in terms of material properties and appearance. Moreover, dense concrete copings are likely to provide better protection to

Fig. 6.27 Homersfield Bridge, Suffolk. The cast iron parapet was repaired and missing sections replaced

the body of the wall than all but the best types of stone. On the other hand the apparent technical advantages are outweighed by the intrusion into a heritage structure of a modern material and concrete copings would only be acceptable on less sensitive bridges.

Maintenance of masonry balustrades may be necessitated by vehicular damage or by erosion and deterioration of the sculpted stone. The latter rarely involves all the balusters as the softer more vulnerable stone deteriorates first. An example of a recently repaired balustrade is shown in Fig. 6.28.

Concrete

As for all concrete structures, the most common mechanism of deterioration in parapets is through corrosion of the steel reinforcement. This presents a conservation problem that has not yet been satisfactorily addressed. The steel reinforcement is invariably very close to the surface of the concrete and sections may be thin, so that conventional patch repairs would be relatively shallow in depth and ineffective in the long term. Repair methods commonly used to date have involved removal of loose and salt-contaminated concrete, cleaning the reinforcement, and making good using a cementitious or polymer-modified mortar. Criteria for selection and formulation of the mortar have been concerned with factors such as workability, adhesion, density, etc., but less attention has been paid to compatibility with the existing concrete and appearance. An example of patch repairs on a precast parapet panel is shown in Fig. 6.29. The fresh repair is an entirely different colour from the host materials and it remains to be seen as to whether it will eventually weather to a more compatible appearance. It also remains

Fig. 6.28 Repairs to stone balustrade. Damaged balusters have been replaced like-for-like

Fig. 6.29 Precast parapet panel having patch repairs. The panel had been in service for about 65 years when repaired. The appearance of the poorly matched repairs has been worsened by chalk marks that have not been cleaned off

to be seen as to whether the repair remains effective, without cracking or spalling.

Deteriorated precast concrete parapets have sometimes been repaired using sprayed concrete. It has the advantage that it is more suitable than patch repairs on thin concrete sections, but its texture and colour change the appearance of old and weathered concrete, as discussed in Chapter 12. An example of an overhead stiffener on a precast concrete footbridge repaired by sprayed concrete is shown in Fig. 6.30.

Replacement of individual members, such as panels and posts, is an altogether more satisfactory strategy as it can produce a lasting and more durable repair. The appearance of the concrete can be more easily matched and in any case small differences are less pronounced when construction joints separate the new and old surfaces. More importantly, there is an opportunity to increase the thickness of cover to the reinforcement or substitute corrosion-resistant reinforcement, such as stainless steel, coated bars or polymers. There is also an opportunity to provide additional strength when this has been found to be required. Unfortunately there are few occasions when this is practical, as concrete tends to deteriorate fairly uniformly and it would be necessary to replace most if not all the members rather than a few of the most damaged ones.

There have been several reported cases where concrete parapets have had to be replaced. Salginatobel Bridge in Switzerland has parapets 1.1 m high and only 100 to 160 mm thick. They are composed of thin concrete panels and pillars having cross-sections of 150 mm × 600 mm. They were designed as non-structural and simply prevent people from falling off. The conservation authorities were very strict about maintaining originality as the bridge is listed as an International Historic Civil Engineering Landmark (as described in Chapter 12). Deteriorated sections of the parapets were replaced to requirements that the original shape, irregular forming and the tight dimensions were reproduced as closely as possible.

Donner Summit Bridge in California, USA, see also Chapter 12, has concrete balustrades which, along with other parts of the structure, had deteriorated over the 70 years service life and exposure to de-icing salt and

Fig. 6.30 Precast concrete repaired by sprayed concrete

weathering. However, the federal authorities no longer permit open balusters and it was necessary to obtain special approval for a compromise that enabled the style if not the detail of the original architecture to be retained. The replacement balustrade was designed with a 25 mm relief and chamfered grooves on the internal face to represent the gaps between the balusters. On the external face, where there are no mandatory restraints, the topography matched more closely the original architecture.

Baltimore Street Bridge across Gwynns Falls in Baltimore, USA, see also Chapter 12, had solid parapets composed of posts at 3 m spacings and panels having an air gap at the bottom so that any impact loading would be resisted by the posts alone. Furthermore, the parapets were widened at intervals to form bases for lamp standards in a manner that was in violation of current federal requirements. The replacement parapet was designed to meet the up to date requirements, but with an appearance as similar as possible to the original. The main revision was to close up the airspace at the bottom so that reinforcement could be added and horizontal impact load shared with the posts. The inside faces of the parapet had reliefs to represent the original air gaps.

Hillhurst Louise Bridge in Calgary, Canada, see also Chapter 12, suffered deterioration over 75 years service due primarily to a combination of freeze–thaw damage and, to a lesser extent, alkali–aggregate reaction. Among other requirements, it was necessary to replace the concrete balustrades with precast units having an appearance similar to the originals. In order to meet current requirements, it was necessary to modify the height and add vertical steel bars between the balusters. The bars were fitted to reduce horizontal gaps to less than 150 mm. They were protected from corrosion by being galvanised and surface blasted to facilitate good adhesion for a finishing coat of black urethane paint. The concrete components were precast individually and prefabricated into 3.66 m lengths. The balusters were connected to the coping and pedestal by grouted rods through their centres. The concrete was treated with a penetrative silane coating and two coats of pigmented sealant. The prefabricated lengths were erected on site between *in situ* cast pilasters. The undersides of the pedestals were shimmed and the gap was filled by grouting after the pedestals had reached sufficient strength.

STRENGTHENING

Strengthening old parapets against vehicular impacts, like conservation and repair, requires special care to minimise change to the appearance of the original structure. It usually requires the addition or substitution of new material which in itself is fraught with problems and potential for spoiling the appearance. Some types of parapet, for example timber, do not easily lend themselves to being strengthened, while others are more suitable. In general, authorities have tended to let well alone unless there is good reason to do otherwise.

The County Surveyor's Society carried out vehicular impact tests on masonry parapets and, from the results, prepared guidelines on design and assessment. Methods of strengthening masonry parapets were investigated.

- Reconstruction in the same form but using higher strength mortar. The effects can be evaluated from charts prepared from the results of the tests.
- Reconstruction to a higher mass, by increased thickness or height. The effects can be evaluated from charts.
- Point the mortar in masonry joints and grout dry stone parapets.

It should be noted that substitution of higher strength mortar for repointing, strengthening dry stone parapets, or reconstruction, reduces the flexibility of the masonry and its ability to cope with thermal movements. In any case the introduction of strong cement mortar into old masonry is not favoured by heritage authorities and introduces various practical problems as outlined in Chapter 7. The method is not applicable to ashlar construction having very thin joints.

Other methods were considered, but no tests or evaluations were carried out.

- Installation of internal reinforcement steel.
- Fix safety barriers to the traffic face of the parapet to spread the impact load and mobilise a greater length of parapet. Lead-in and lead-out lengths would be required for the approaches.

It follows that strengthening masonry parapets has to be a compromise between minimising changes to their appearance and achieving high containment strength.

While it is fairly easy to demonstrate that an old parapet fails to meet modern performance requirements it can be difficult to show that a new design or a retrospectively strengthened parapet complies. The problem arises from the dynamic response because the parapet should not only be strong enough to contain the errant vehicle, but do it in a manner that minimises the danger to both occupants of the vehicle and to other road users. Sophisticated numerical modelling can be carried out, but the most convincing method is to carry out a crash test on a full-scale model albeit this is usually seen as being too expensive to justify for a single parapet. One of the ways round this problem is to incorporate a standard barrier that has been tested and approved.

Some of the methods used to strengthen old parapets are outlined in the following sections and summarised in Table 6.2.

Substitute stronger materials

Ornate cast iron parapets have considerable aesthetic appeal and are invariably heritage listed. However, the materials used at the time had very low fracture toughness and therefore little resistance to impact from errant vehicles. Also, the ironwork invariably had weak connections to the deck so that it had low resistance to crowd loading let along vehicle impact. One of the strengthening measures that can be taken is to substitute a tougher material.

On Westminster Bridge, London, the brittle cast iron parapets were

Table 6.2. Methods of strengthening parapets

Method of strengthening	Advantages	Disadvantages
Substitute higher strength mortar in masonry parapets	An economic method straightforward to carry out.	Liable to crack under thermal stress, (may be necessary to introduce expansion joints on longer parapets). Cement mortar joints can lead to damage in old masonry. Unacceptable on historic bridges.
Substitute stronger materials, e.g. ductile cast iron	Can be designed to match appearance of original. Acceptable to heritage authorities.	Loss of original materials. Expensive.
Internal reinforcement of cast iron.	Can be designed to incorporate standard barriers. Little change in appearance. Acceptable to heritage authorities.	Some loss of originality.
Internal reinforcement of masonry	No change in appearance.	Some applications can be expensive.
Strengthen connections to deck.	A common weakness that can usually be tackled without changing appearance of the parapet and deck.	Disruptive to traffic and expensive.
Additional barrier between pavement and road	Standard barrier meets impact requirements. Appearance can be designed to be compatible with the bridge. Provides protection to pedestrians. No change to outside appearance of bridge. Economic.	Disruptive to traffic and expensive.

replaced with spheroidal graphite cast iron, grade 400/18 L20. The new casting was designed to have the same external appearance as the original, but with improved connections between the 1.95 m long units and the addition of internal stiffening. The parapet was connected to a new reinforced concrete deck by stainless steel bolts. This provided an altogether stronger connection.

Table 6.2. Continued

Method of strengthening	Advantages	Disadvantages
Safety kerbs	Installation causes minimal interaction with existing structure. Provides protection to pedestrians. Can be designed to have an appearance compatible with rest of the bridge. Acceptable to heritage authorities. Economic.	Degree of protection less than for a standard barrier.
Construct standard barriers	A straightforward design that provides the required level of impact strength.	Loss of originality. Changes appearance of the bridge and can be particularly unsightly when P6 concrete barrier erected. Unacceptable to heritage authorities.

Special attention was given to the paint as it was desired to return the bridge to its original colour scheme. To this end, paint samples were removed and analysed and it was found that the colour above deck had been pale green. This was reproduced on the new parapets and lamp standards.

Internal reinforcement

Strengthening by introducing internal reinforcement can be carried out by reconstructing the whole parapet or inserting reinforcement into the existing parapet. Naturally, reconstruction is less acceptable to heritage authorities, but in some cases there is no alternative. This is exemplified by North Bridge, Edinburgh constructed in 1896 and having under-strength cast iron parapets by modern standards. The parapets were composed of 3.66 m long panels bolted to the steel deck at mid-points and to cast iron posts at their ends. It was found that severe corrosion had occurred in the steel deck. Also, the cast iron had an unfavourable chemical composition making it brittle to impact loading. In consequence, the calculated strength under lateral loading of the parapet was very low. After consideration of several schemes it was decided to erect a P1 barrier along the parapet line, enclosed by spheroidal graphite cast iron units detailed to match the original parapets. By agreement with the rail authority, the height was maintained unchanged at 1.4 m, but the width had to be increased to accommodate the 100 mm wide longitudinal members of the P1 barriers. The profile of the traffic-facing sides of the parapets was modified to reduce projecting details from 50 mm to 25 mm to comply with regulations. The concrete deck was extended beneath the new parapet to provide protection for the steelwork and seatings for the P1 barrier. The ductile cast iron parapets were estimated as having lower first cost and lower maintenance replacement costs than a GRP alternative scheme. In the final analysis the parapet work cost 65% of the total cost of renovating the bridge.

On Findhorn Bridge in Scotland, a standard barrier was erected with GRP external facings detailed to be compatible with the original wrought iron parapets. Where items are attached to the outside of a parapet for visual effect care must be taken with the detailing so that on impact pieces do not shatter, fly off and cause damage.

On Magdalene Bridge, Cambridge, the cast iron parapet has been strengthened by fitting prestressing strand longitudinally tensioned between the pilasters at each end. The pilasters have reinforced concrete cores and are clad in stone to give an appearance similar to the originals. The prestressing strand is located at the top of the parapet and at the bottom where it is partially exposed, as shown in Fig. 6.31. This is an excellent scheme as the strengthening has made no change to the appearance of the bridge and the original ironwork has been retained.

In a recent development masonry parapets can be strengthened by insertion of vertical reinforcement grouted in a sock to contain leakage through the mortar joints and achieve structural continuity with the masonry. Transverse strength is achieved by introduction of horizontal bars beneath the road surface. This method is applicable to masonry arch bridges. A variant of the method has been designed for strengthening stone balustrades against crowd loading. This incorporates vertical reinforcement and longitudinal reinforcement through the copings.

Fig. 6.31 Magdalene Bridge, Cambridge. The parapets have been strengthened by strands fitted at top and bottom, and tensioned between the pilasters

Additional barrier

In cases where it is not found to be feasible to strengthen parapets by other means, and there is sufficient available space, freestanding crash barriers can be erected at the boundary between road and pavement. Although this is generally regarded as being unsightly and spoiling the view of the parapet from the road, it has the advantage of providing protection for pedestrians and accidental wheel loading on under-strength footways. It also leaves the old parapet in its original and unspoiled condition and the external profile of the bridge is unchanged. Examples include Queen's Avenue Bridge, Aldershot, where the original parapet of ornate cast iron had been refurbished, as shown in Fig. 6.32, Farnham Road Bridge, Guildford, which had brick parapets, and Woodbridge Old, Guildford, which had reinforced concrete parapets.

On Gala Water Bridge in Scotland, a standard barrier was erected with imitation Georgian iron railings fitted to the outer facing side. This provided an external appearance compatible with the bridge, but the view from passing vehicles was of course modified.

As an alternative to the installation of a standard barrier, Battersea Bridge across the Thames in London, has an 800 mm high 'pedestrian guard', designed to be compatible with the Victorian decoration on the existing parapets and on the rest of the bridge. In addition dished kerbs have been installed, as shown in Fig. 6.33.

On a narrow bridge, standard barriers would have an obtrusive appearance. However, the effect can be lightened by a modified design such as that used for Battersea Bridge.

Fig. 6.32 Queen's Avenue Bridge, Aldershot. A freestanding standard barrier provides protection for pedestrians and leaves the refurbished cast iron parapet in original condition

Fig. 6.33 Battersea Bridge, London. Pedestrian guards designed to be compatible with the Victoriana and dished kerbs provide protection against errant vehicles

Safety kerbs

Profiled safety kerbs of increased height provide a restraint to errant vehicles albeit not as effectively as a standard barrier. The effectiveness is dependent on the speed of vehicle, angle of approach and whether road conditions are wet or dry. The Trief Kerb is of precast concrete, 324 mm high and has an edge profile designed to provide frictional resistance which decelerates an impacting vehicle. Trief Kerbs have been installed on a number of bridges in Sandwell and on Battersea Bridge, London.

On Cleveland Bridge, Bath, the cast iron parapets were found to be inadequately connected to the deck and under strength. An initial scheme to replace the original cast iron was shelved in view of the historic value of the parapets which had been constructed in 1827. The revised scheme incorporated a high safety kerb in cast iron. The units were designed to be compatible with the parapets and had vertical fluting to relieve their profile and special details at each end to make a positive statement in harmony with the lodges, where tolls were formerly collected, see Fig. 6.34. The parapets were provided with improved connection to the deck so that all the posts would act as cantilevers and resist lateral loading. Broken and cracked components were repaired or replaced.

Construct standard barrier

The special requirements for parapets on bridges over railways are difficult to meet satisfactorily, particularly when the parapets are of masonry. Fig. 6.35 shows a masonry arch bridge having standard concrete P6 parapets installed in place of the original brickwork. This has been done without any effort to harmonise with the existing structure and has resulted in a ghastly

Fig. 6.34 Cleveland Bridge, Bath. The original cast iron parapets of 1827 have been repaired and retained. Cast iron safety kerbs have been installed to restrain errant vehicles

Fig. 6.35 Masonry parapets replaced by P6 standard concrete barrier. This has resulted in a ghastly mismatch of materials and proportions

mismatch of materials and proportions. The railway beneath the bridge has subsequently been closed so that the parapet has become even more out of context.

BIBLIOGRAPHY

Guidance Note for the Assessment and Design of unreinforced Masonry Vehicle Parapets (1995). County Surveyors' Society.

Chapter 7

Masonry bridges

GLOSSARY

Intrados	The lower face of the arch barrel
Extrados	The upper face of the arch barrel
Springing	The plane, usually extra strong, on which the ends of the arch sit
Spandrel	The longitudinal edge wall, which sits on the barrel and retains the fill
Arch barrel	The main structural element of an arch bridge
Corbel	Structural element projecting from arch face (usually to support widening)
Pattress plate	Load-spreading plate fitted at ends of tie-bars to restrain spandrels
Spreader beam	Load spreading strip over the length of the span and fitted at ends of tie-bars to restrain the spandrels.
Voussoirs	Wedge-shaped stones that together form the arch barrel
Anchors	Grouted steel reinforcement bars
Gothic arch	Arch having a pointed shape
Masonry	Bricks or stones
Starling	Piles, usually timber, driven into river bed upstream of bridge pier to provide protection against floating material.
Cutwater	Wedge-shaped abutment to piers to divide the water flow and provide protection against floating material and water erosion.
Clapper bridge	Primitive river crossing composed of flat stone slabs laid across raised stone piers.
Packhorse bridge	Masonry arch bridge on packhorse route, typically no wider than 2 m and constructed mainly in the period 1650 to 1800.

BACKGROUND

There are some 40,000 masonry arch bridges in Britain on roads, railways and canals and an unknown number of clapper bridges. The oldest surviving arch bridges date back to before the fourteenth century and there are many fine examples still in use. The early bridges had relatively short spans, were often multispan, and typically constructed with random masonry, see Fig. 7.1 for example. There were many fine river bridges built in the eighteenth and nineteenth century exemplified by the work of, for example, Smeaton, Telford, Rennie and Harrison. They were a departure from the earlier rudimentary bridges, as they had longer spans and were often embellished with architectural features.

Over the years masonry arch bridges have experienced various changes as they have been successively widened and strengthened so that many are now far from original. This must be taken into account when planning conservation work.

Many of the early bridges had no parapets, or parapets that were inadequate to meet modern containment requirements. Various methods of strengthening and reconstruction have been used, many resulting in loss of originality and an ugly appearance as described in more detail in Chapter 6.

Clapper bridges have received less 'modernisation' as they are generally in remote locations and only used by pedestrians and animals. Nevertheless, in some cases, rather ugly railings have been installed. There are a few clapper bridges big enough to carry vehicular traffic, for example Park End Bridge in North Yorkshire and Pencarrow Bridge in Cornwall, shown in Fig. 7.2.

As mentioned in Chapter 4, there are some 189 packhorse bridges identified in England. In addition, there are many others that have been widened so that they are no longer easily recognisable.

Fig. 7.1 Barton Bridge. An early fourteenth century bridge of 4 m width. Due to being located on an unmetalled route not used by vehicles, it has remained in original condition

Fig. 7.2   Pencarrow Bridge. A clapper bridge carrying vehicular traffic

Some notable masonry bridges are listed in Table 7.4 at the end of this chapter.

The impact of utilities

It is important to be aware of the impact of utilities (gas, electricity, water and telephone). The attachment of utility pipes has had an aesthetically and in some cases structurally damaging effect on all bridges and masonry arches in particular. When the utilities have been attached externally they have invariably spoiled the appearance of the side elevation, see for example Figs 7.3 and 7.4. When the utilities have been buried in the structure the trenching may have been through medieval fill causing damage to material of archaeological significance. More importantly, on occasions when the depth of fill over the crown has been insufficient to contain the utility pipe, the arch barrel has been damaged by being notched to provide the necessary depth, often without leaving any external evidence. This weakens the arch and presents a difficult situation, on occasions when it is necessary to strengthen the bridge to raise its load carrying capacity.

There can be other side-effects caused by utilities, for example Huntingdon Bridge, a grade I listed structure, was damaged when a water main burst over the arch barrels. In general, maintenance work on utilities involving trenching damages the waterproofing (if present) and can lead to leakage and deterioration of the masonry.

Information about utilities is rarely recorded with any accuracy and weakened structures can go unnoticed. On other occasions, utilities have been taken through the arch in the vicinity of the springing so that the side elevation is spoiled, clearance beneath the arch is reduced and the structure is weakened, see for example Fig. 7.5.

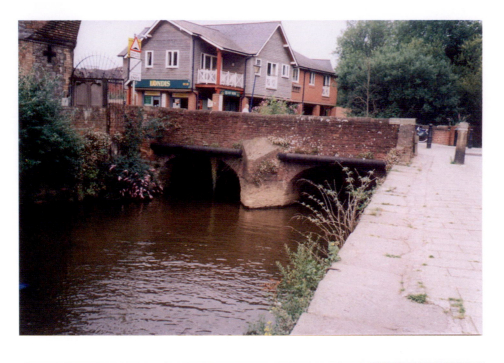

Fig. 7.3   External attachment. Utility pipes located beside the arch of this bridge producing an unsightly appearance made worse by the mortar-covered extension above the cutwater

Fig. 7.4   Attachment to parapets. Services fixed to the inner face of the parapet producing an unsightly appearance and the need for strengthening

Performance

Masonry bridges are a very durable form of construction, requiring comparatively modest levels of maintenance. There have been few structural failures and those that have occurred have been mainly through storm damage when swollen and fast running rivers have caused substructures to fail

Fig. 7.5 Internal attachment. Utility pipes positioned through the arch causing structural weakening, reducing the headroom and restricting the watercourse

(more detail is given in the section on substructures, later in this chapter). There have been occasions when spandrels have peeled off due to development of transverse stresses. Normally, failing spandrels give ample warning so that corrective action, such as fitting transverse ties, can be taken. Kettlewell Bridge in North Yorkshire was an exception. It was known to require repairs and was being monitored when a 6 m length of spandrel wall and parapet fell off without warning in February 1985. It was believed to have been caused by a freeze–thaw cycle when wet fill expanded and pushed off the spandrel.

There have been instances of disused masonry arch bridges collapsing without warning. This has been when they have been 'out of sight out of mind', and there have been no inspections and no maintenance. Mortar can leach out of joints and thermal cycles can generate upward movements during hot spells that are not corrected by the compressive action of live loading. These actions can result in a progressive distortion and 'loosening' of the arch structure until the point is reached when the masonry falls to pieces. This occurred in a disused bridge in Nottinghamshire in the 1980s and Norris bridge across the Basingstoke canal at Fleet in 1979. In both cases the bridges were of brick construction. There is an implication that historic bridges that have been down-rated to pedestrian use, or set aside altogether, cannot be left to the elements and assumed to be maintenance-free, but require continued surveillance to ensure that they remain serviceable and safe.

There are very few masonry arch bridges that are less than 100 years old, and most are much older. Although designed to carry horse-drawn vehicles, they are coping well with the repetitive action of heavy goods vehicles of up to 40 tonne gross weight. Fatigue is seen as being a potential problem, but has not been positively identified as occurring to date. However, this could change as the numbers of cycles of heavy axle loads progressively accumulate.

CONSERVATION

Masonry bridges have been managed and maintained using a range of techniques, some meeting the requirements of conservation, some not. As with all types of bridge, the better known ones have been looked after with care and the masonry has been maintained using techniques developed for historic buildings. The lesser bridges have been treated differently as the objectives of highway authorities have been to ensure that they meet the requirements of contemporary traffic within limited budgets. Methods of conservation are described in the following section and summarised in Table 7.1.

Repoint mortar joints

Repointing of the mortar joints is an essential maintenance activity which, if left undone, can result in damage by weathering and loss of strength. Masonry bridges were usually constructed using lime-based mortar and it is best to repoint with the same material. Whereas a lime-based mortar can deform and permit expansion and contraction with changes in ambient temperature, cement mortar is likely to crack in an unsightly manner which may cause damage to the masonry. Lime mortar is porous and permits moisture to exit from the structure. In contrast, cement mortar keeps moisture in so that material behind the surface can become saturated and susceptible to frost damage. In earlier times, when the ductility requirement was not properly appreciated, cement mortar tended to be used for repointing and there was consequential damage to the stones which became sacrificial in place of the mortar. It should be added that using small quantities of cement to achieve a quicker set is also unsatisfactory. The importance of using lime mortar has become recognised in recent times and on occasions when maintenance is being done, the opportunity can be taken to replace earlier cement mortar repairs. There is excellent guidance on lime mortars and techniques of repointing available in specialist publications (see Ashurst, 1988).

It is also important to ensure the mortar is correctly profiled. If it is smeared across the stonework, water can penetrate behind the edge of the smear and subsequent weathering can force off the mortar, as shown in Fig. 7.6. This can lead to damaged stonework.

Replacement masonry

With the deterioration in air quality in recent years, acid rain and the use of de-icing salt during cold weather, masonry has been eroding and degrading much more rapidly than in the past. When the damage becomes excessive, it is necessary to take action, but the situation should be properly studied beforehand to identify the cause of the damage. For example, it may be

Table 7.1. Conservation of arch bridges

Method of conservation	Advantages	Disadvantages
Repoint mortar joints	An essential job required to maintain strength, conserve the stonework and exclude water.	There are no disadvantages, but the correct material must be used. Incorrect mortar can be unsightly. Excessively strong mortar can lead to unsightly cracking and damage to the masonry. Lime mortar takes longer to cure.
Masonry replacement	Durable method of repair having a pleasing appearance.	Incorrectly matched masonry can be unsightly and weather differently. Correct (like-for-like) masonry can be more expensive.
Coatings	An effective method of protecting friable stonework. No change in appearance.	Requires a full understanding of the situation. Coatings have a finite life and become ineffective after about 15 years. Expensive.
Sprayed mortar	A quick and economic method of repairing defective masonry faces.	The material is not fully compatible with masonry. The hard mortar can crack and spall off. Appearance can be unsightly.
Cementitious patches	A quick and economic method of repairing local defects. Careful mix design can ensure a good colour match so that the appearance is unchanged. Economic.	The material is not fully compatible with masonry. The harder patch can spall off the masonry.
Anchor debonded arch rings	Rapid and economic. No significant change in appearance.	No obvious disadvantage if carried out properly.
Anchor cracked barrels	Rapid and economic. No significant change in appearance.	No obvious disadvantage if carried out properly and provided the cause of cracking has been correctly identified. Some cracking is evidence of articulation and best left alone.

Table 7.1. Continued

Method of conservation	Advantages	Disadvantages
External reinforcement of barrels and rings	No obvious advantages over more modern methods now available.	A rather cumbersome method no longer used. Changes the appearance. Expensive.
Lateral restraint of spandrels	A traditional and well proven method. Relatively cheap and effective making minimal change to the appearance.	Tie-bars may corrode and fail if subjected to chlorides but this risk may be exaggerated (stainless steel is sometimes used).
Correct distorted arches by partial reconstruction	A comprehensive approach that restores structure to its original condition. No change in appearance.	Causes disruption to traffic. Expensive.
Stabilise distorted arches by anchors	Minimal disruption to traffic. No change in appearance. Economic.	No obvious disadvantages if carried out properly.
Reconstruction	Enables a grossly deteriorated structure to be fully restored plus opportunity to raise load carrying capacity.	No obvious disadvantage if carried out properly but loses the patina of age and originality. Likely to be more expensive than construction of a modern replacement.

due to leaking water in which case waterproofing may be required. Repair is usually by cutting out the affected blocks and replacing them. It is preferable, in every way, to replace like-for-like using the same type of stone or brick. It is rarely possible to obtain stone from the original quarry, but stone of the same geological type can be substituted. Examples of good conservation are provided by the repairs to the Avoncliff Aqueduct on the Kennet and Avon Canal, see Fig. 7.7, and the limestone pavilions on Ouse Valley Rail Viaduct at Balcombe. Here it was found that it was more cost-effective to use the existing stone, with new stone where required, than a complete rebuild using reconstituted stone.

In contrast, there are numerous examples where incompatible materials have been substituted, for example ill-matched bricks in dressed stone faces in Fig. 7.8.

The issues involved in the maintenance of stonework apply equally to brickwork, possibly more so, because brick tends to be treated with less respect and there are numerous examples where repairs have been made using ill-matched bricks, see for example Figs 7.14 and 7.15. Obtaining the correct replacement brick can require the production of hand-made products. This was done for the Balcombe Viaduct (mentioned above) when repairs were made to the arches and piers using hand-made bricks. Moreover the arch repairs were carried out using traditional arch-turning

Fig. 7.6 Incorrect pointing. Mortar has been smeared over the edges of the masonry and weathering has subsequently loosened it

Fig. 7.7 Avoncliffe Aqueduct. Defective ashlar blocks replaced by stone of the same type to make a seamless repair that will weather to the same colour and texture as the rest of the structure. The inset shows repairs in progress

methods. Conservation of one of the bridges across the Basingstoke Canal in Hampshire, involved the production of hand-made coping bricks for the parapets. This was all the more meritorious as it was by volunteers and funded by subscription.

Penetrant coatings

On occasions when the surfaces of stonework are deteriorating and eroding, penetrative coatings can be used to stabilise the situation. Suitable coatings

Fig. 7.8 Dissimilar materials. Several types of brick used to repair dressed stone at different times giving a patchwork appearance made worse by calcareous leakage

have been well researched and there is considerable experience of using them to protect historic masonry buildings, statues and other structures over many years. There are two types of penetrant coating; hydrophobic and consolidant. Hydrophobic coatings line the pores within the masonry and prevent water from entering, but permit water vapour to exit. As a secondary benefit, the masonry is modestly strengthened to the depth of penetration. In contrast, consolidant coatings strengthen the masonry and provide a small degree of hydrophobicity. For soft stone or brick having a porous structure, coatings can penetrate to depths of as much as 60 mm. There is a range of well-developed coatings available commercially. They can be applied by spray or brush and are colourless so there is no change to the appearance of stonework. However, they can be expensive and their use tends to be limited to parts of heritage structures and valuable architectural details. Coatings are fairly commonly used on masonry in other European countries, but much less in Britain. Penetrative coatings have not been generally applied to masonry bridges and on occasions when surfaces are deteriorating it is usually considered more satisfactory to replace the masonry. There is, however, a case for using coatings to protect architectural details when they are becoming eroded, provided due care is taken to diagnose the cause of the damage. It is necessary to recognise that, if used in the wrong circumstances, they can worsen the situation. For example, it is not uncommon for salty water to soak into the masonry from the roadway during winter months and remain in the pore structure. Under these circumstances, the application of coatings can trap the salt and subsequent build-up can lead to surface layers of masonry spalling off. Also coatings have a finite life of about 15 years before they become ineffective and require renewal. At present there are no guidelines on how to assess the coatings and decide when they have become ineffective.

Sprayed mortars

Damaged masonry can be repaired using sprayed mortar, sometimes supported by reinforcement mesh. The process is outlined in the Conservation section of Chapter 12. The process provides a strong cover of high density which is resistive to any further environmental damage. It was originally developed for repairing concrete and is well suited to this purpose as it is a compatible material. In the as-sprayed state it can have a roughcast appearance and a colour and texture different to most types of concrete and masonry. In consequence, it is best applied to parts of the bridge that are not in the normal sightlines. Sprayed mortar is less suitable for application to softer and more porous types of sandstone and limestone, because the interfacial bond between the dissimilar materials is more likely to fail. Care has to be taken to ensure that there is no leakage behind the repair otherwise there is a risk that the hard skin will crack and fall off. Sprayed mortar can be unsightly and is not now accepted for heritage structures. Examples of sprayed mortar are given in Figs 7.9 and 7.10.

Fig. 7.9 Sprayed mortar on to brickwork. The contrast between the appearance of the brickwork and mortar is lessened by adjacent foliage. The mortar is supported on a steel mesh

Fig. 7.10 Sprayed mortar on to stonework. A typical example where the mortar has cracked and developed a dismal appearance. The mortar is unreinforced

Cementitious patches

Locally damaged masonry is sometimes repaired using cementitious or polymer-modified materials. A cementitious repair material may typically be composed of cement of suitable colour, lime and stone dust proportioned to provide adequate strength and a good colour match. It is essential to prepare the base material to provide good mechanical keying and a surface free of any friable material. As with sprayed mortar, there is a risk of creating a hard and incompatible coating which will eventually fall off.

Although the repair material can be formulated to have the same appearance as the masonry, this can change when the structure is wet, particularly if a polymer-modified formulation is used. Also, the repair material is likely to weather to a different appearance.

Anchor debonded arch rings

Arch ring separation is fairly common with multi-ring brick arches, an example is shown in Fig. 7.11. Laboratory tests on large-scale models have shown that complete debonding of the rings can reduce the load carrying capacity by 20 to 30%. In practice, debonding is unlikely to be total but, for the purpose of assessment, has to be assumed unless there is good reason

Fig. 7.11 Cracking indicating debonded arch ring. Debonded arch ring and bulging spandrel. Note the pattresses and waling timbers fitted to this heavily trafficked urban bridge

to do otherwise. A more realistic assessment can be carried out if data can be obtained by coring through the masonry and noting the areas of debonding. Radial anchors can be installed to repair ring separation and movement between the arch barrel and spandrel walls. Installation is by diamond core drilling through the masonry and grouting a reinforcing bar, preferably stainless steel, in place. It is preferable if not essential to use a sock to contain the anchor and grout, so that the injection pressures do not displace or push out bricks. On completion, any defective pointing should be repaired and the entry point for the anchor made good. In places where appearance is particularly important, a core containing the external surface of the masonry should be retained so that it can be bonded into a recess over the anchor. If the original core is used there is likely to be an unsightly annulus containing bonding materials between the core and parent masonry. It is therefore preferable to install a core having a tighter fit and use bonding material having a compatible appearance. Alternatively new pieces of compatible masonry may be fitted.

It may also be necessary to fill voids between the arch rings using an appropriate grout. The grout should be formulated to have suitable strength, an adequate setting time to enable the various activities to be carried out, and good flowability. Injection should be after the anchors have been installed and any repointing has been done, otherwise there is a risk that even modest injection pressures may cause grout to leak through cracks and spoil the appearance of the masonry.

Anchor arch barrel cracking

It is not uncommon for the arch barrel to crack in a span-wise direction beneath the spandrels, see Fig. 7.12. This may be due to a variety of causes,

Fig. 7.12 Cracked arch barrel due to transverse deflection of spandrels

for example live loading causing a flexible arch barrel to deflect in a vertical plane relative to a less flexible spandrel wall. A common method of dealing with cracked arch barrels is to stitch across the crack with short tie-bars or anchors.

The construction activity is the same as for installation of anchors to stitch together ring separation. However, if the flexibility of the arch barrel is left unchanged, the problem may reappear elsewhere. It follows that it is prudent to add stiffness to the barrel. This can be achieved by methods such as reinforcing the barrel, or reducing the stresses transmitted to the barrel by increasing the thickness of the surfacing.

However, the cause of the cracking should always be investigated and properly diagnosed before proceeding as stitching should not be considered an automatic solution. In some cases the cracking is evidence of articulation and best left alone.

Arch barrels have sometimes been stabilised by external reinforcement. In the example shown in Fig. 7.13 steel rails have been used to strengthen the barrel in a lateral direction and steel plates have been positioned against the face of the barrel to act as pattresses and distribute any lateral stress that may be generated. This method was occasionally used in the early part of the twentieth century. In the example shown in Fig. 7.13, its appearance

Fig. 7.13 Externally reinforced arch barrel. This bridge has a brick barrel and the rest of the structure is dressed stone. Some of the bricks have been replaced with ill-matching ones of a different colour. Steel rails have been fitted to strengthen the barrel in a lateral direction

has weathered to become compatible with the bridge but the effect has been spoiled by subsequent repairs using ill-matched bricks.

Lateral restraint of spandrels

Spandrels can sometimes be displaced laterally in an outward direction. This can be due to the action of live loading developing lateral thrust through the fill. The displacement may be due to outward rotation, transverse sliding on the arch barrel, circumferential cracking of the arch barrel, or bulging. In the past, the most common method of dealing with this has been installation of tie-bars through the width of the bridge to restrain any further movement. The tie-bars are secured at each end by nuts and washers over load-spreading pattress plates. The pattresses are in different shapes: crosses, sigmoids or dishes. Variants such as spreader beams that run the length of the span have been used. Over the years these have weathered rather well and are accepted as adornments that are in harmony with the masonry, see Figs 7.14 and 7.15. Cast iron pattress plates are preferred to steel, as they are less likely to produce unsightly rust staining. In the time since the introduction of de-icing salts in cold spells, tie-bars have become less popular as it is feared that they may corrode. It is now more likely that the opportunity would be taken to strengthen the structure by methods such as placing concrete behind the spandrels to reinforce them locally, excavate the fill and replace by low strength or foamed concrete, or inject grout into the fill at strategic locations. However, tie-bars and pattress plates are preferred by heritage authorities. The fear of corrosion may be an over reaction to the experiences with reinforced and post-tensioned concrete. In any case the preferred strategy should be to ensure that water is kept out of the structure.

Fig. 7.14 Pattress plate. Bricks in the arch ring have been replaced by a different type and the spandrel has bulged so that the remedial pattress plate is bedded on to different levels

Fig. 7.15 Spreader beam. The parapet has been damaged and repaired on several occasions using different types of brick

Correct distorted arches

The curvature of the arch barrel can sometimes become distorted for one reason or another, see Fig. 7.16. This can be the start of an escalating situation because the running surface is likely to be deformed causing vehicles to impose impact loads on to the bridge as they cross it.

Fig. 7.16 Distorted arch. Local view showing cracks and ring separation

Skirfa Beck Bridge in North Yorkshire, had a severely distorted arch with outward separation and displaced voussoir stones as shown in Fig. 7.17. After consideration of alternative methods, fill was removed from the deformed side and the arch was jacked back into the correct profile, the

Fig. 7.17 Skirfa Beck Bridge. Correction to the distorted profile using timber trestles and hydraulic jacks

masonry was repointed, a concrete saddle was cast on to the arch barrel and the fill was replaced. Use of rapid-setting cement in the concrete enabled high early strength to be obtained so that the time of closure to traffic was minimised. This achieved a good repair that retained the original material and appearance while enhancing the strength of the bridge at modest cost.

In an extreme case, Prestwood Bridge, a brick arch across the Staffordshire and Worcestershire canal, constructed by Brindley circa 1770 had fallen into a state of disrepair and the arch had become so badly distorted and weakened that it was closed to traffic. The parapets became unsound and were removed almost to the level of the road surface. Part of the spandrel had bulged by 100 mm in relation to the arch barrel. It was decided to rebuild the bridge as closely as possible to its original form. As part of the demolition, TRL was invited to load the bridge to collapse. Line loading was applied at quarter-span where the arch was deformed downwards and therefore most vulnerable to additional load. The bridge exhibited a surprisingly high level of remnant strength and eventually failed at 22.8 tonne. (The assessed strength using the Military Engineering Experimental Establishment (MEXE) method was a permissible axle load of 2 tonne, the predicted collapse load using a sophisticated mechanism analysis was 17.3 tonne.) It follows that distorted arches may not be as weak as they sometimes appear albeit this is not a margin of strength that can always be assumed. The bridge was subsequently reconstructed to be as close as possible to its original form.

Reconstruction

Partial or total reconstruction is an expensive activity and is usually confined to bridges of special interest. Moreover it is not favoured by heritage authorities, except as a last resort when the alternative is replacement by a modern structure. On occasions when reconstruction is carried out, it should incorporate as much of the original structure and material as possible so that the appearance and texture is unchanged. At the same time, those bridges carrying normal traffic should be improved to the extent that they have the capacity to carry full traffic loading.

Laigh Milton Viaduct in Ayrshire is a good example of careful and correct reconstruction. It is the oldest surviving public railway viaduct having first carried traffic in 1811 and is therefore of special historic interest. It had been neglected for many years and reached the stage of imminent collapse when it was partially reconstructed and re-opened in 1996. The original masonry was re-used and where this was not possible, new masonry of similar type was introduced. Lime mortar was used in the reconstruction. A 0.3-m hogged distortion was replicated in one of the arches. The spandrels were infilled with mass concrete to waterproofing level. Undercut piers were secured by reinforced concrete collars.

Tilford Bridge in Surrey was constructed in the thirteenth century and is a listed structure. Two of the arches were reconstructed in 1998. Care was taken to use original techniques; the stonework was random ironstone and a prescribed lime mortar was used. There was an added complication during autumnal weather when it was necessary to erect a canvas tent and install heaters to aid the curing of the lime mortar. During the work, a seventh arch was discovered. Archaeologists carried out a watching brief during the dismantling phase but nothing of significance was discovered. A view of the reconstructed bridge showing the cantilevered wooden parapet is shown in Fig. 7.18.

Prestwood Bridge (mentioned earlier) was demolished and reconstructed using as much of the original material as possible. The reconstructed arch was of a higher strength having a barrel composed of a single-course thickness of brick which acted as facing to a reinforced concrete saddle. Parapets were reinforced by stainless steel bars. Concrete fill was topped by a waterproofing membrane. The external appearance was faithfully reproduced including the distorted shape of the arch. However, heritage authorities do not require reproduction of defects that have developed in the time since original construction, nevertheless it is a thoughtful touch.

Mytchett Place Bridge on the Basingstoke Canal, constructed by Pinkerton circa 1790, was originally a typical brick arch. In the early 1900s, the barrel was replaced by simply supported rolled steel joists and mass concrete infill. The facades were left unchanged but the abutments were extended above the springing by engineering blocks to support the deck, see Fig. 7.19. When inspected in 1992, it was found that this rather nondescript structure had deteriorated badly and a 3 tonne weight restriction had to be imposed. In view of the historic interest in the bridge, and its position on a site of special scientific interest, it was decided to reconstruct it as a replica of the original design, but about 1.5 m wider and able to carry full traffic loading. Where it was necessary to dismantle the masonry, the stones were carefully logged and reinstated in the same relative positions. The reconstructed bridge, shown in Fig. 7.20, was composed of reinforced concrete faced by the masonry. Interestingly, blast furnace slag cement was used in the con-

Fig. 7.18 Tilford Bridge, Surrey. The bridge was refurbished and two of the arches were painstakingly reconstructed in 1998. The footway is cantilevered off the structure

Fig. 7.19 Mytchett Place Bridge before reconstruction. Rolled steel joists fitted in the early 1900s can be seen projecting below the crown of the original arch

Fig. 7.20 Mytchett Place Bridge after reconstruction. Great care was taken to reproduce the facade in its original appearance

crete to reduce the early thermal cracking and plastic shrinkage. The concrete was reinforced with polypropylene fibres, rather than steel, in order to reduce the possibility of the concrete attracting increased bending stress to the masonry.

WIDENING

The early masonry bridges were rather narrow and by the eighteenth century the more intensively used structures had to be widened. Tymsill Bridge, a single span structure in Bedfordshire, provides an interesting and fairly typical history of successive widening. The original fifteenth century bridge was 2.3 m wide and built in rubble limestone. During the seventeenth and eighteenth centuries, it was widened on the upstream side, using limestone, to 3.4 m and later to 4.6 m. In the nineteenth century, it was widened on the downstream side, using limestone abutments and a brick arch, to 5.8 m. In 1937, it was widened on both sides, using concrete, to 10 m. Regrettably most of the medieval construction is now out of sight and the appearance of the bridge is dominated by the final widening in concrete.

Widening presents aesthetic problems. When the same profile and appearance as the original structure is adopted, the aspect ratio is reduced and the bridge loses some of its character. Also, a gloomy tunnel effect can be created beneath the bridge when the span and clearance are relatively short. When the widening is of a different structural form, great care has to be taken to avoid a clash in styles. Methods of widening are summarised in Table 7.2.

Table 7.2. Methods of widening arch bridges

Method of widening	Advantages	Disadvantages
Connected structure of same profile	Causes minimum change leaving one side in its original state.	One side of the original structure is covered up and lost from view. Can create a dark and unfriendly tunnel effect beneath the intrados.
Connected structure of different profile	Usually the cheapest solution. Can be designed so that some of the original structure remains in view.	Presents aesthetic problems, can be successful or can be positively ugly.
Separated structure	The original bridge is unchanged. Sensitive designs can produce a pleasing effect.	Insensitive designs can create an ugly clash in styles.
Cantilevered structure	A cheap solution suitable for footpaths.	Generally unsuitable for heritage structures. Changes appearance of side elevation.
Over-decking	Can be successful structurally and aesthetically. Successfully applied to convert railway structures to highways.	Completely changes the appearance of the deck. Aesthetically not suitable for structures of low height.

Contiguous structures

The early widening schemes usually involved construction of an adjacent structure having matching spans and butted against the original bridge leaving stonework of the spandrels in place. Widening was not always carried out in sympathy with the original structure, for example the fourteenth century nine-span arch bridge at Bradford-on-Avon had two Gothic arches and the rest semi-circular. When the bridge was widened in the seventeenth century with arches of matching spans, all were semi-circular so that two of them butted against Gothic arches creating an unnecessary clash in styles. Design of such a mismatch against a 300-year-old bridge would almost certainly attract adverse criticism in present times. However, the new part was constructed in the same type of stone and has weathered over the years causing the mismatch to become quaint and acceptable rather than ugly. In fact, the original bridge probably had nine Gothic arches of which seven were reconstructed at some earlier and unrecorded date so that the widening followed the same strategy. In contrast, Barton Bridge located about 800 m downstream has survived in its original state with four Gothic arches, see Fig. 7.1.

Stokeford Bridge, a four-span structure across the River Avon, was widened in 1929 from 3.2 m to 7.6 m and again in 1964 to 9.5 m between parapets. The first widening was in mass and reinforced concrete having all visible areas faced in the same type of stone and to the exact detail as the original. The second widening incorporated curved beams and was also faced in stone to the same detail. An overslab was laid over the whole area, so that there was no potential for leaking connections. The bridge was sufficiently tall and long to be widened without changing its proportions to an unacceptable extent. The materials have weathered well and the connection between new and old parts has become seamless.

Widening schemes have usually been designed with little or no structural connection between the adjacent structures. In consequence, longitudinal cracking has sometimes developed at the interface. This is usually cosmetic and best left alone, because the two halves may have different stiffness under traffic loading so that structural connections could develop high local stresses leading to cracking and failure. In cases where the longitudinal cracking is reflected in the running surface or permits leakage through the structure, it may be appropriate to take other action such as overslabbing. On occasions when there is good reason to have a structural connection, it can be provided by techniques such as insertion of transverse anchors to tie together the new and old arch barrels.

Widening in more recent times has sometimes been less pleasing. In the 1930s, reinforced concrete became in vogue and a typical widening scheme of this era is shown in Fig. 7.21. Here a concrete portal extension has been cast against a nineteenth century stone arch. The differing materials and geometries have created an ugly and unnecessary clash in styles in a scenic rural location. Whereas the stone extension at Bradford-on-Avon had mellowed with time, 60 years' weathering of the concrete has, if anything, worsened it to a scruffy 'downtown' appearance.

There are cases where ancient bridges were widened by the construction of contiguous structures and subsequently restored to their original state.

Fig. 7.21 Clash of styles. An example of widening where different materials and geometries have been adopted

For example, Rotherham Chantry Bridge, dating from 1483, was widened in 1768. In 1921, parts of the structure were found to be defective and it was decided to build a new bridge a few metres upstream and the work was carried out in 1928–1929. The additional structure was removed from the ancient chantry bridge and the parapets reconstructed on the original line using old stone and new coping.

Marton Bridge in Warwickshire, dating from 1414, is the oldest bridge on a trunk road (but due to be detrunked). The bridge is 80 m long having two river arches of 4.2 m span and two 2.7 m flood arches. It originally had a width of 4 m with no footways. In 1926 it was widened to 13 m by the addition of a reinforced concrete portal span of 9.7 m framing the two main arches but concealing the rest of the structure, see Fig. 7.22. The original ashlar stone parapet was transferred to the new concrete structure. In the 1990s, the concrete structure was found to be inadequate to carry 40 tonne vehicles and it was decided to construct a new vehicular bridge a few metres to the north. In 1999–2000, the concrete structure was dismantled to reveal original stonework on the north side of the ancient bridge, see Fig. 7.23. Unfortunately some of the stone had been damaged by the strong cement mortar used in 1926, and had to be replaced. The medieval mortar was chemically analysed so that a similar mix could be used in the reconstruction. The parapet was carefully dismantled and rebuilt on the original line. Archaeologists were present during the work, to record features and artefacts as they were revealed. The ancient bridge is now restricted to pedestrians. This is an excellent example of the restoration of an ancient monument.

Fig. 7.22 Marton Bridge, Warwickshire. Appearance of 1926 north elevation. Two of the fifteenth century arches could just be seen through the rectangular opening (but not in this view)

Fig. 7.23 Marton Bridge, Warwickshire. Appearance of the medieval north elevation immediately after removal of the 1926 concrete widening

Separated structures

In a number of cases, arch bridges have been indirectly widened by construction of a separated structure to take traffic in one direction, leaving the original bridge to take traffic in the other direction. This is more satisfactory from a heritage viewpoint as it leaves the original bridge unspoiled. However, care is required to ensure an aesthetically pleasing combination of styles. New structures can be designed to be slimmer so that some of the original arch can still be seen. This effect can be more successful for multi-arch bridges, because the new structure can have fewer spans so that there is a relatively wide and clear space beneath its deck. Nevertheless the view of that side of the original bridge is inevitably spoiled.

A particularly elegant scheme was designed for a structure adjacent to Runnymede Bridge across the Thames. The original bridge had been designed by Lutyens and was located in an exceptionally sensitive location. The new bridge was designed as an open concrete arch having prestressed concrete frames at each end. The profile was the same as the original and there was a 3 m gap between the structures. The view of Lutyens' bridge was obscured as little as possible and space and light were preserved.

Other examples of separated structures include Elstead Bridge, a thirteenth century bridge across the River Wey in Surrey. The modern structure has two spans and is designed to leave as much of the original bridge in view as possible, Fig. 7.24.

Cantilevered structures

Widening by cantilevered additions can enable the original width of the bridge to be used by vehicles and the new part by pedestrians. It can be on one side or both sides of the bridge depending on requirements. An example of the former is Tilford Bridge in Surrey, shown in Fig. 7.18.

Newton Cap Bridge, Bishop Auckland, a grade I listed structure built in 1388, originally had a carriageway of only 4 m width. In 1900, it was widened

Fig. 7.24   Elstead Bridge, Surrey. An adjacent structure has been designed to leave as much of the original thirteenth century bridge in view as possible. The inset shows the view from above with the original structure on the right hand side

Fig. 7.25 Newton Cap Bridge. A fourteenth century structure widened in 1900

to 4.5 m by reducing the width of the parapet walls, and cantilevered footways of 1.24 m width were constructed on either side as shown in Fig. 7.25.

An aesthetically pleasing example of cantilever widening is shown in Fig. 7.26. Here a multispan brick and stone arch bridge has been widened

Fig. 7.26 Middle Bridge, Newport Pagnell. An example of the use of similar materials to achieve a pleasing effect

on one side using cast iron corbels to support springing points for the new structure. The other side has been left in its original condition. By raising the height of the additional arches, the tunnel effect has been lessened. The new brickwork is of a similar mellow appearance to the original and an interesting reflective effect has been achieved. Surprisingly, the brick arch ring of the addition is laid differently having three, separate rings of headers, whereas the old arch has the same thickness, but with English bond where headers and stretchers are laid so that it is a single structural component.

A less pleasing example of cantilever widening was carried out on Wye Bridge in Kent. Here a seventeenth century five-span grade II listed stone bridge was widened in the nineteenth century. Recesses for pedestrians were removed and footpaths attached on either side by iron frames fixed to the sides of the bridge. Writing in 1930, Jervoise described the widening as having 'completely spoilt the appearance of the bridge'. In addition the bridge has had utility pipes attached to the masonry which are unsightly and would be expensive to remove.

Over-decks

Widening can be achieved by constructing a new deck on top of the existing arched structure. This was carried out in 1995 on Newton Cap Rail Viaduct, Bishop Auckland, a grade II listed structure built in the 1850s, to achieve a pleasing and practical conversion from rail to road, see Fig. 7.27. The existing width of 7.3 m was increased to 11.3 m between parapets and the new deck was 450 mm thick and 257 m long, continuous over the full length of the viaduct. In order to minimise the additional weight of concrete, transverse reinforced concrete support walls at 5.2 m centres were constructed mono-lithic with reinforced concrete saddles over the arches. The transverse walls

Fig. 7.27 Newton Cap Rail Viaduct. In 1995, the structure was converted from rail to road to relieve traffic on the historic bridge nearby

support the continuous deck on steel sliding bearings. The deck was fixed near the middle of the viaduct and there were guided expansion joints at other piers and abutments. The level of the deck was raised by 1 m, which permitted more of the existing masonry to be retained and created 2.3 m headroom within the interior for ease of access and inspection. Footways were located on the cantilevered sides of the new deck. The retained arch structure was repaired; defective bricks in the intrados of the barrels were replaced, damaged sandstone was repaired with colour-matched mortar and areas beneath the piers and abutment were probed and grouted.

A similar type of over-decking scheme was proposed for the Wye Bridge in Kent, but eventually turned down partly because it was seen as being too wide for the proportions of the old bridge. Cantilevered sides are not considered an attractive option for widening smaller arch structures, as the straight edge of the cantilever and the shadow effect detract from the original disposition of the masonry. On longer and taller bridges, however, with careful detailing and retention of the original parapet, cantilevering can be successful.

Bray Viaduct, a structure supported on tall masonry piers, was widened from a single-track rail to a road having a 7.3 m carriageway, in 1989. Considerable care was taken to achieve a pleasing appearance having minimal environmental impact. To this end a new concrete deck was constructed having its edges raised to reduce transmitted traffic noise and conceal the view of traffic from the valley below. The conversion involved raising the height of the six existing piers and installing three others to replace an existing and inadequate tipped embankment. This new construction was faced with matching local sandstone and gritstone. The increased height and length of the viaduct enhanced the already dramatic effect on the local terrain, see Fig. 7.28.

STRENGTHENING

Masonry arches are intrinsically strong and durable, but it is often difficult to demonstrate their full strength analytically. Despite having been around for thousands of years, there has been an inadequate understanding of their full structural action. When assessing load carrying capacities, it has been necessary to use approximate methods to obtain safe values for all types and geometries of arch. As a result, many arch bridges have been unnecessarily strengthened or replaced. The real strengths were demonstrated vividly in a series of tests to collapse carried out by the Transport Research Laboratory between 1985 and 1989 (see Page, 1993). Live loading was applied at quarter or third points to give the most damaging configuration. Collapse loads for seven tests were between 108 and 560 tonne and an eighth bridge which was in a very advanced state of deterioration and closed to all traffic, was collapsed at 22.8 tonne. In the time since these tests, more realistic methods of analysis have been developed and tested against the TRL data so that it has become possible to identify under-strength arches more accurately. To this end, the discrete element method has been developed to provide a particularly powerful analysis of masonry behaviour and ultimate strength.

Fig. 7.28 Bray Viaduct.
An elegant conversion
and widening of a single-
track rail bridge to a
two-lane highway bridge

Various methods of strengthening superstructures have been carried out as described in the following sections and summarised in Table 7.3.

Identification of hidden strength

Before preparing a strengthening scheme for a masonry arch bridge it is prudent to recheck maintenance records to ensure that strengthening has not been carried out at some time in the past. It is not unknown for arches to have been strengthened by reinforced concrete saddles, overslabs or other schemes which have been recorded but overlooked, or not recorded and only discovered after new work has been commenced. These types of strengthening are not superficially apparent but can easily be investigated by taking cores at strategically selected locations.

Structures in their as-built state, with no strengthening, can appear to have inadequate load carrying capacity according to conventional methods of assessment, but have hidden reserves of strength. This hidden strength can be identified using more comprehensive methods of analysis. Before carrying out the desk work, however, the structure should be briefly investigated to determine whether there are any signs of distress, such as cracking. If there are any doubts about dimensions, cores should be removed to confirm factors such as the actual thickness of the arch barrel and whether there is extra backing above the springing points. Analysis using the method of discrete elements enables more accurate modelling of the blocks of masonry, mortar and the fill. This method has been used on numerous occasions on bridges where it has been possible to demonstrate that strengthening is not required.

Table 7.3. Methods of strengthening arch bridges

Method of strengthening	Advantages	Disadvantages
Identify hidden strength	Enables continued use without strengthening.	None
Saddle	An effective method of increasing strength. Provides opportunities for archaeological studies and waterproofing. No change in appearance.	Difficult if utilities are present. Extensive disruption to traffic. Expensive if temporary bridge and temporary propping are involved. Maintenance problems may occur due to reinforcement corrosion.
Overslab	No change in appearance.	Possible difficulties if utilities are present. Disruption to traffic. Maintenance problems may occur due to reinforcement corrosion.
Overbuilding	A rapid method of reconstruction.	Loss of originality and an ugly clash in styles.
Thickened surfacing	A fairly effective method. Simple to carry out. No change in appearance. Economic.	Limited increase in strength. Effects difficult to calculate with precision as surfacing material is variable, degrades and softens in warm weather.
Prefabricated liner • Corrugated steel • GRC panels • Reinforced concrete	An effective method of increasing strength. No problems with utilities. Can be relatively easy and quick to erect.	Disruption to traffic if across road or railway. Headroom reduced. Maintenance problems may occur due to steel corrosion. Unsightly and unsuitable for heritage and other sensitive structures.
Supportive truss	An effective method of increasing strength. No problems with utilities.	Disruption to traffic if across road or railway. Reduced headroom and unsuitable for many situations. Maintenance work likely to be required. Unsightly and unsuitable for heritage and other sensitive structures.

Analysis using discrete elements has the added advantage that it can identify less common modes of failure such as shear and is not confined to the hinge mechanism.

Table 7.3. Continued

Method of strengthening	Advantages	Disadvantages
Sprayed liner	Can be relatively easy and quick to carry out.	Disruption to traffic if across road or railway. Headroom reduced. Maintenance problems may occur if reinforced. Unsightly and unsuitable for heritage and other sensitive structures.
Internally reinforced arch barrel	An effective method of increasing strength. Minimal disruption to traffic. Stainless steel unlikely to require any maintenance. No change in appearance.	Care has to be taken to avoid damage to utilities. Accurate drilling required to avoid breaking out of the arch barrel.
Externally reinforced arch barrel	A quick method of increasing strength.	Changes appearance of soffit somewhat.
Replace fill with weak concrete	An effective method of increasing strength. No change in appearance.	Possible difficulties if utilities are present.
Grout the fill with flowable cementitious material	A limited method of increasing strength. No change in appearance.	Possible difficulties if utilities are present Unsuitable for some types of fill. Grout quantities are difficult to predict.

Saddle

The arch barrel can be thickened and strengthened by adding concrete to the extrados. The concrete, which is usually reinforced, forms a saddle. The construction process requires the fill to be excavated and the arch barrel to be exposed and cleaned. It is important for the concrete to form a structural connection with the masonry so that the full potential strength can be developed. It has been estimated that, depending on the circumstances, a 33% increase in barrel thickness could double the load carrying capacity. Circumstances that dictate the degree of strengthening include: strength of the concrete, whether the concrete is reinforced, degree of composite action and relative stiffness of the concrete and masonry.

Saddling has been a popular method as it is seen as being straightforward and direct, and the opportunity can be taken to waterproof the arch. However, there can be difficulties if utilities are present, particularly when there is little depth of fill over the crown of the arch. With the increasing densities of traffic and expectation of the travelling public, road or lane closures for the duration of saddling, have become less practical. Also, it is necessary to take care to ensure that the spandrel walls are stable when the fill has been removed and there is no risk of collapse.

Overslab

Construction of an overslab enables axle loading to be distributed more evenly through the fill and into the arch barrel. It also provides an opportunity to waterproof the bridge and properly manage the run-off. In designing the overslab, however, it is necessary to ensure there is no local overloading of the barrel as could be caused if there is an inadequate thickness of fill over the crown. One method of ensuring that such problems cannot occur is to build up the abutments so that a suitably designed overslab can be simply supported at each end to act as the load-carrying element. The arch barrel is left supporting the fill and is effectively redundant.

As with saddling, the presence of utilities in masonry arches may add complications and construction operations requiring road closures, which may not be acceptable.

In one of the few examples of a clapper bridge required to carry vehicular traffic, Park End Bridge in the North York National Park was found to be seriously understrength. Several stone slabs in the 13 m long seven-span structure were cracked. The stones were overlaid by a reinforced concrete slab cast on to a polystyrene layer. This method preserved both the originality and the appearance of this interesting and unusual bridge.

Overbuild

Overbuilding is a technique carried out mainly in the late 1800s and early 1900s, whereby a steel or composite superstructure was constructed on to the masonry abutments in place of the arch. This could be designed to raise the load-carrying capacity and increase clearance beneath the deck, as shown in Fig. 7.29. Although an economic and rapid method of reconstruc-

Fig. 7.29 Example of overbuilding. An ugly mismatch of styles

tion, it usually resulted in an unattractive clash of styles and has been described as hideous. Its appearance is particularly awful when it is applied to one or two arches in a multispan viaduct leaving the rest unchanged. In Mytchett Place Bridge, the change in appearance was minimised by leaving the spandrel walls unaltered, but the steel beams could be seen projecting below the crown of the arch, see Fig. 7.19. It was very creditable that Surrey County Council chose to remove the beams and reconstruct the bridge in its original form.

Thickened surfacing

Thickening the surfacing has been advocated as an economic method of increasing load carrying capacity by distributing axle loading more evenly. However, it is unlikely to be as effective as overslabbing of the same thickness, as asphalt is less stiff than concrete and becomes more flexible in warm weather.

It is necessary to pay due attention to the longitudinal surface profile to even out the 'hump' of the arch, if one is present, and arrange for the added thickness of surfacing to be tapered down to some distance from the bridge to avoid a surface step. When vehicles travel over surface irregularities, impact factors are generated which can result in amplified loading of the arch. Surface thickening can accompany other strengthening methods, so that the additive effects produce the required results. On its own, thickening the surface is likely to be a less potent and less reliable method than most others and the degree of strengthening is difficult to calculate with any precision. Moreover, as surfacing wears and cracking develops its structural contribution is lessened.

Prefabricated liners

Arches can be strengthened by fitting prefabricated liners beneath the barrel. Various materials have been used including corrugated steel, fabricated sheet steel, glass reinforced cement and reinforced concrete. It is necessary for the liner to follow the arch profile reasonable accurately so that the resulting space between liner and arch can be filled with grout to achieve a structural connection.

A prefabricated liner changes the appearance of the soffit and side elevation of the arch. The liner can be sculptured to become an architectural feature otherwise it is liable to be positively ugly. Liners of any type reduce headroom beneath the arch.

When the three-span Spencer River Bridge, Northampton, was found to be understrength and in need of widening, various schemes were studied. It was decided that the best option was to line the arches with precast segmental units supported on corbels, bolted to the piers and abutments. Design was based on the safe assumption that the barrels had failed and made no contribution to strength. The annulus between the brick arch and liner was grouted with PFA cement grout. The abutments were underpinned with mini-piles from road level down to stiff clay. Care was taken with the appearance and the new elevation was faced with blue engineering bricks to match the existing upstream elevation. The side faces of the liners were in-stepped relative to the brick arch-rings to soften the effect of the concrete.

Supportive truss

A supportive truss is a variant on a prefabricated liner. Whereas liners have a curved profile which follows that of the arch and minimises loss of headroom, supportive trusses change the profile to 'square off' the curvature. It follows that the headroom is significantly reduced and the method is only suitable for tall arches where there is headroom to spare. Supportive trusses can be constructed in steel or concrete, for example Fig. 7.30. They produce a stark contrast in material and engineering style which is usually unattractive and unsuitable except in a last resort.

Sprayed liners

Masonry arches can be strengthened *in situ* by external reinforcement and sprayed concrete, to form an *in situ* liner. This is distinct from the use of sprayed concrete to repair defective masonry or brick. The concrete is projected at high velocity so that a high density and strength can be achieved. It is reinforced by steel to provide the added strength. The spraying process requires experience and skill to ensure concrete fills the crevices and holes

Fig. 7.30 Supportive truss. A structurally effective method, but having a discordant contrast in materials and engineering style that has not softened with time

that are likely to be present in the masonry. There is also the 'shadow effect' to be avoided when spraying around steel reinforcement. The process can apply concrete up to 300 mm thick. Care must be taken to ensure that there is a good interfacial joint, otherwise the sprayed lining may become detached from the masonry due to shrinkage or deterioration of the masonry surface.

As mentioned earlier in the section on conservation, the appearance of sprayed concrete can be rather poor and is unsuitable for heritage structures. Special attention is required for the edge detailing to avoid an excessively unsightly appearance in side elevation. Sprayed liners reduce headroom beneath the arch.

Internally reinforced barrel

Internal reinforcement of arch barrels by grouted anchors has been developed commercially as a proprietary system. The system is designed to meet individual requirements of the geometry, location and condition of the bridge in question. The anchors comprise stainless steel reinforcement bars positioned in a sock and grouted *in situ*. The grout expands the sock to fill any cavities or holes within the masonry, but prevents undue pressure from building up and displacing bricks or stones. Sufficient grout can leak through the sock and provide structural adhesion against the masonry. The anchors are installed in holes drilled perpendicular to the axis of the barrel and contained within the barrel tangential to the curvature. The drilling can be done from above or below the arch depending on ease of access, presence of utilities and traffic, see Fig. 7.31. The method of strengthening is very efficient as relatively few reinforcement bars are usually required and their disposition can be designed to suit the masonry. Under normal conditions, the site works can be carried out very quickly and with little interruption to traffic. There is no change in appearance and the method is favoured for heritage structures. In fact, the method is regarded as one of the

Fig. 7.31 Internal reinforcement. Holes drilled from the surface of the deck prior to insertion and grouting of reinforcement anchors

least intrusive ways to raise load carrying capacity. It has been used to strengthen historic masonry arch bridges in Britain, Australia and the USA.

Externally reinforced barrel

Methods of externally reinforcing arch barrels have been developed commercially whereby longitudinal and transverse slots are cut in the intrados and reinforcement is embedded in a suitable resin mortar. The systems also include embedment of radial ties in the arch. One of the commercial systems employs 6 mm diameter stainless steel reinforcement, the other employs a more flexible steel reinforcement.

During the retrofitting work, the arch is structurally weakened and therefore vulnerable in the time between cutting the slots and the mortar curing to full strength. Depending on the design and geometry of the arch this weakening may, or may not, be structurally significant. In some cases there may be a need to take precautions such as closing the bridge to traffic, limiting the number of slots cut at a given time and carrying out the strengthening on successive areas of the intrados, or the arch might have to be propped.

External reinforcement changes the appearance of the intrados somewhat, albeit this can be minimised by chasing out mortar joints to provide slots for the reinforcement and by careful workmanship. In any case, most of the surface of the soffits being affected is usually in the shade and not easily seen.

Replace fill with concrete

There have been occasions when strengthening has been achieved by replacing the fill with concrete. This effectively converts the structure to masonry clad mass concrete. Depending on the geometry, the concrete can be of relatively low strength, or it may require local reinforcement. The latter is a variation on the more common method of saddling the arch and topping-up with mass concrete. Replacing the fill has the same disadvantages of requiring closure to traffic until the concrete has cured to a sufficient strength, and it is unlikely to be feasible if utilities are present.

Grouting the fill

As an alternative to mass concrete, it may be feasible to inject grout into the fill. This is of course dependent on the nature of the fill, for example it is not feasible if there is a high content of fines. It is necessary to ensure that pointing is in good condition otherwise grout can leak out to cause unsightly staining and environmental pollution. The grout must be flowable, stable and to have a setting time of 1 h or more. The 7-day cube strength should be around 10 N/mm^2.

SUBSTRUCTURES

Substructures and resistance to flooding could be regarded as the Achilles heel of masonry arch bridges, as flood damage is by far the most common cause of failure. In an appraisal of 143 failures of all types of bridges up to 1976, it was found that sixty-six were due to scour and many of the bridges were masonry arches. In the time since this appraisal, there have been more failures caused by scour, for example the 127-year-old, five-span brick arch railway bridge across the river at Inverness collapsed in a 1-in-100 year flood in 1989. Here, the failure was attributed to scour beneath the piers due to a record river flow of 800 m^3/s.

The older masonry arch bridges, particularly those having comparatively short spans, are not very efficient in a hydraulic sense because the channels between abutments and piers are invariably a serious constraint to the flow of water during floods. The constraints cause water levels to rise on the upstream side, so that hydraulic forces are imposed on the structure. Also, fast flowing water can scour the bed of the river and undermine the foundations, for example a river bed could be scoured to a depth of 600 mm, whereas foundations may be only 300 mm deep. There are various ways to improve the hydraulic performance of bridges without significantly spoiling their appearance.

Hydraulic efficiency

Hydraulic efficiency can be improved by the provision of entrainment or guide walls to funnel the flow approaching and leaving the bridge and provide a streamline transition from the river bank to the bridge and vice versa. Entrainment walls should preferably be constructed with large stones or gabions. Concrete or sheet piling spoils the setting of the bridge and damages the habitat and wildlife. The shape of piers in plan can be made less bluff by rounding corners, this has usually been done in the past by the provision of cutwaters on the upstream side and sometimes on both sides.

A more recent scheme has been to construct suitably shaped guides, composed of piles in a diamond shape (in plan) upstream of the bridge, in an attempt to control scour. This has been done for several railway bridges on a semi-experimental basis. The guides are treated as sacrificial because scour is relocated away from the bridge piers.

A more traditional method of protecting the piers is by the provision of starlings to a height just above low-water level.

Scour can be controlled by the provision of an invert (inverted arch) having its concave face uppermost and located in the river bed between the toes of abutments or piers. Early inverts were constructed of masonry, whereas modern ones are concrete. Scour can occur beneath the invert on the downstream side. This can be controlled by installation of trench sheeting. Inverts can also be constructed as flat slabs, sometimes referred to as paving. There are early examples having historic surfaces, such as cobbles and stone pitching and these should be retained when possible. Nowadays concrete is commonly used but tends to settle and crack with time. Heritage authorities favour the use of gabions for both paving and retaining the riverbank. In addition to protecting from scour, inverts act as struts that oppose inward movement of abutments, as for the famous Ironbridge in Shropshire.

It has been suggested that inverts can be designed to improve hydraulic efficiency by sloping them downwards in the convergent section of flow approaching the abutments, and sloping them upwards in the divergent section. This compensates for the reduced channel width, enabling water to flow through the arches at a more uniform speed so that scour is minimised.

Some of the older multispan river bridges have so-called flood arches. These are normally dry but provide extra hydraulic capacity when water levels rise during floods. It is necessary to attend to these during maintenance to prevent them becoming blocked by debris, landfill or new construction.

Conservation

Where there is a need to repair or enhance foundations, underpinning can be carried out by excavating material and replacing it with mass concrete, or by insertion of minipiles. The method of underpinning and its design is of course influenced by the nature of the ground and the strengthening requirements. Minipiles are typically installed so that they run through the pier or abutment, usually at an angle inclined to the vertical, and into the ground beneath. They make a direct connection between the substructure and suitable foundation ground beneath. Additional works to make a structural connection, such as installation of a pile cap, are not required.

Weak piers are sometimes strengthened by steel frames, a method that is effective but changes the appearance of the bridge and is unattractive.

Undercut piers can be strengthened by reinforced concrete collars resting on suitable foundations, as for Laigh Milton Viaduct, Ayrshire.

MANAGEMENT PRACTICE

Inspection

Management of masonry bridges is invariably hampered by a shortage of information about the internal features and dimensions, and the maintenance history. This shortfall can be a problem with all types of bridge, but is more prevalent in masonry bridges because of their greater age and chequered history. Factors that can be important, if not crucial, to the assessed strength include:

- the actual thickness of the arch barrel, as opposed to the apparent thickness indicated on the end-face, which may be different
- dimensions of backing above the springing, if present
- dimensions of any retrospective strengthening that may have been carried out such as a reinforced concrete saddle or overslab
- details of the design of any additions that may have been made to the original bridge. The most common addition is to widen the bridge, usually by building an adjacent structure having the same profile
- dimensions of paving between abutments, if present, and

- location of utilities, if present. Accurate positioning is required if it is proposed to carry out strengthening work from above the deck.

When this information is not available, it will be necessary to obtain it by physical measurement. Suitably located cores can be removed to determine thickness of the arch barrel, dimensions of backing, dimensions of a saddle and dimensions of an overslab.

Widening is usually recorded on file, but in any case can be identified from the appearance of the intrados. It may have been carried out on one side or both sides of the bridge. Information about whether the new and old parts are structurally connected are rarely available and it is reasonable to assume there is no connection. When working on the bridge, allowance should be made for the possibility that the original spandrel (on the side that has been widened) may have been left in place. On the older bridges this is historic material that should not be damaged.

The identification of paving may require an underwater inspection.

The recorded locations of utilities are often insufficiently precise so that it may be necessary to dig a small trench through the road surfacing to obtain adequate information to enable some types of strengthening work to proceed. Magnetic detection is a non-invasive method that can be used with reasonable accuracy.

The methods of inspection most commonly used to determine the condition of the structure are:

- visual to detect the presence of cracking, leakage, distortion, state of the mortar pointing, etc.
- impact echo (hammer tapping) to detect voids and weak or deteriorating masonry
- radar to detect internal defects such as voids, the presence of utilities and internal geometry of the structure
- magnetic detection of the presence of ferrous materials such as reinforcing bars, and
- underwater inspection of the substructure to detect the presence of scour, etc. This is an important issue as scour is by far the most common cause of collapse of masonry arch bridges.

Maintenance

Masonry structures are very durable but there are several types of defect than can develop over the years and require maintenance:

- Bulging spandrels can be caused by lateral pressures in the fill. They are usually stabilised by fitting transverse ties and pattress plates (see Figs 7.14 and 7.15), preferably cast iron as explained in the section on distorted arches. If the movement is left unchecked, the spandrels may eventually fall off. Alternatively, the movement may leave voids that can fill with water, which freezes in cold spells and generates additional lateral pressure which may be repeated each winter.
- Longitudinal cracking in the arch barrel reflecting the position of the

spandrel above, see Fig. 7.12. Some maintenance authorities have considered it necessary to remake the transverse structural connection in the arch using dowel bars and making good the mortar. However, current thinking is that the cracking is a result of traffic loading causing the arch structure to articulate separately from the spandrels and as such it is harmless and best left alone unless there is progressive widening of the crack.

- Longitudinal cracking between the new and old parts of a widened structure. This is usually best left alone unless there is good reason to impose a structural connection.
- Thermal cracking of masonry where the fill has been replaced by concrete. There have been occasions when this has occurred and it is necessary to carry out a structural analysis of the situation before designing an appropriate solution.
- Leakage is self-evident and usually due to failure of the waterproofing. Sometimes weep holes have been installed in the spandrels to permit the water to escape. Usually, the waterproofing can be made good at the appropriate time in the maintenance cycle. Unchecked leakage can permit erosion of particles from the fill, the generation of voids and weakening of the structure.
- Ponding of water and consequential damage can be caused by accumulated detritus. Gulleys, passageways, weep holes, gargoyles and drains should be kept clear. Waterproofing of the decks should be maintained in effective condition.
- Leakage from water pipes buried in the arch has sometimes occurred causing considerable damage, as mentioned earlier. Early action should be taken if leakage becomes evident.
- Cracking in the spandrels can, among other things, be due to unequal settlement of foundations or the action of live loading. Masonry structures can cope with a degree of distortion caused by unequal settlement and it will often be sufficient to repoint with lime mortar. If the movement is still active, it will require investigation and action to stabilise it. Cracking, due to the action of live loading, is more serious as it can lead to the development of a collapse mechanism, albeit arch bridges can continue to carry load when in an apparently very advanced stage of failure, as evidenced by Prestwood Bridge.
- As mentioned earlier in this chapter, serious damage is sometimes caused, apparently unnoticed, by installation of utilities. This can present some difficult problems as utilities can be very expensive to move and it is usually necessary to work around them.
- Deterioration of the jointing mortar requires attention and repointing from time to time. It is important to use lime mortar, as cement mortar is incompatible with old masonry and can cause damage. The pointing must be carried out so that the mortar has the correct profile.
- Deterioration of individual pieces of stonework or bricks can reach the stage when they must be replaced on a like-for-like basis.
- Vegetation should be removed and kept clear of the masonry.

BIBLIOGRAPHY

Page, J. (1993) *Masonry Arch Bridges*, TRL State-of-the-Art Review, HMSO.

Jervoise, E. (1930) *The Ancient Bridges of the South of England*, The Architectural Press.

Public Works, Roads and Transport Congress. (1933) *British Bridges*.

Highways Agency. (1996) *The Appearance of Bridges and Other Highway Structures*, London: HMSO.

Ashurst, J. (1988) *Plasters, Mortars and Renders*, Practical Building Conservation Series, Vol. 3. Gower Technical Press.

Waterway Environment Services. (1999) *Design Manual Volume Two. Repair and Conservation*. British Waterways.

Table 7.4 Some examples of masonry bridges

Bridge	*Date*	*Span(s) m*	*Comments*
Tarr Steps, Somerset	BC	17-span 40 m long	Origin and age unknown. Damaged by floods in 1952. Supports specially protected from erosion in 2000. A record of the bridge is kept to aid repairs when stones are displaced.
Monument House Culvert, London	200 AD	0.65	A Roman structure in pristine condition. Now infilled.
Marton, Warwickshire	1414	4 arches	The oldest bridge on a trunk road, now limited to pedestrians. A reinforced concrete widening added in 1926 was removed in 2000 and the original bridge was restored.
Barton, Wiltshire	14th century	4 arches	A typical example having four Gothic arches and in its original condition.
Prior Park, Bath	1755	3 arches	A typical example of an ornamental bridge in the Palladian style.

Dean Valley, Edinburgh	1832	4 arches of 29 m span	Magnificent viaduct 30 m high and towering above Dean Village, by Thomas Telford.
Grosvenor, Chester	1832	61 m	The fourth largest masonry span in the world.
Royal Border Rail Viaduct	1847	28 arches of 19 m span	Stone piers with brick arches and built on a curve, by Robert Stephenson.

Chapter 8

Timber

ABBREVIATIONS

CCA   Chromated Copper Arsenate, a commonly used preservative
EPA   Environmental Protection Act (1990)

GLOSSARY

Anisotropy	Different physical properties in different directions
Cleft (or hewn)	Timber that has been split along the grain, i.e. not sawn
Hygroscopic	Tendency to absorb moisture
Splash zone	Zone at ground level (approximately 200 mm) liable to harsher wetting conditions
Tracheids	Timber cells having a 'straw-type' structure
Boron compounds	Environmentally friendly preservatives containing boron salts
Sister member	A new member fitted in parallel to strengthen an existing member

BACKGROUND

The use of timber as a bridge building material has to a large extent been superseded by steel and reinforced concrete. Worldwide, the stock of timber bridges has fallen significantly in recent years, for example, between 1980 and 1995 the number of bridges in Finland fell from 1200 to 700 and there are similar reductions in other countries. Nevertheless, many bridges still remain, both in and out of service, that are constructed either fully or partly of timber. Timber remains well suited to country park bridges where it fits in with the ambience, see for example Fig. 8.1. Increasing concern over the sustainability issues raised by the use of steel and reinforced concrete has been, at least partly, responsible for a revival of interest in the use of timber for the construction of new bridges, albeit primarily for lightly loaded shorter spans, for example footbridges having spans of up to about 40 m. Unfortunately this change in attitude has not been reflected in an increased interest in the conservation of the older timber bridges.

Timber was frequently used in a secondary structural role in older bridges, one such example of this is illustrated in Fig. 8.2 showing the underside of Union Bridge, Horncliffe, a suspension bridge with a timber deck.

Due to the organic origin of timber and its consequent tendency to decay, timber bridges tend to require more maintenance and more frequent inspections than steel or concrete. The *in situ* remedial treatment of timber is covered later in this chapter, although there will frequently be occasions where the only appropriate response is to replace a deteriorating member. A timber bridge's maintenance programme should therefore anticipate the periodic requirement for replacement of members.

Through careful design and detailing, and programmed maintenance it should be possible to maintain a timber bridge for as long as required. Some notable examples of surviving timber bridges are given in Table 8.7 at the end of this chapter. Although several of these bridges have been rebuilt to their original design, they are all more than 100 years old and demonstrate the sustainability of timber construction. Moreover, the use of modern preservatives and design detailing improves the service lives of new construction and maintenance lives of repairs to old construction.

The examples of surviving timber bridges include Old Shoreham Bridge in West Sussex, a beam-and-trestle structural originally built in 1781, shown

Fig. 8.1 Small footbridge in a country park. The original members will have been replaced possibly more than once

Fig. 8.2 Union Bridge, Horncliffe. Underside of the timber deck showing harmless staining and transverse metal ties

Fig. 8.3 Old Shoreham Bridge, West Sussex. Originally built in 1781 of oak, rebuilt in 1916 using blue gum. It was closed to traffic in 1968 and restricted to pedestrians

in Fig. 8.3 and the well-known Mathematical Bridge in Cambridge originating in 1784 and shown in Fig. 8.4.

An interesting bridge was built across the Thames at Walton in 1751 to the design of William Etheridge. The bridge comprised three arches, with a central span of 39 m, constructed from straight timber members, supported

Fig. 8.4 Mathematical Bridge, Queens College, Cambridge, 1784

on what would almost certainly have been stone or brick abutments and piers. This bridge was recorded on canvas by Canaletto, shortly after construction. The illustration in Fig. 8.5 is based on his painting. Many different theories have been developed regarding the design of the bridge, but what is clear is that a key element of the design is that any one of the wooden members could be replaced without disturbing the others. Sadly this bridge lasted no more than 30 years.

Brunel made extensive and successful use of timber in the construction of railway viaducts but it became clear that rising labour costs, coupled with an overall drop in the quality and availability of suitable timber made their continued maintenance uneconomic. All of Brunel's timber bridges were replaced by 1934.

The Royal Albert Bridge at Saltash, also by Brunel, is an example of the use of timber to aid articulation. The wrought iron structure was originally supported on oak bearings. Over time, as a result of inadequate drainage, the oak began to rot. As part of a recent maintenance programme, the decision was taken to replace the bearings. After careful consideration of the effects on the rest of the structure of possible replacement bearing systems, it was decided that the most suitable material was still timber and

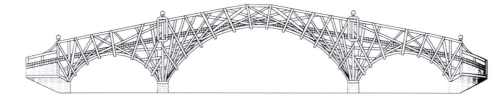

Fig. 8.5 Old Walton Bridge circa 1751. When constructed the central span of 39 m was the longest in England

the rotten bearings were replaced with new oak bearings, identical to the originals that had lasted for some 130 years.

Those timber bridges built prior to 1960, which are still in existence, will invariably be beam, arch, trestle, truss or a combination of these structural forms. Timber has also been used for various components in other types of bridge, for example, decks of suspended and movable bridges, parapets of masonry arch bridges and others, and towers of suspended footbridges.

Timber design guidance is constantly being produced and standards revised to reflect developments in the technology of timber assessment, grading and treatment. These are, of course, relevant to conservation.

Timber properties

Timber possesses lower strength in both tension and compression than steel, and lower strength in compression than reinforced concrete (reinforced concrete is not usually designed to act in pure tension). This means that to support a given load a timber section has to be significantly more substantial than a steel or reinforced concrete section. Timber also has a much lower elastic modulus than steel or concrete (reinforced concrete *per se* does not have a unique elastic modulus). This means that, for example, a beam of a given section supporting a given load will deflect much more if it is made of timber than if it is made of steel or concrete. As a result of this combination of low strength and low stiffness, timber is seldom used for modern structures required to carry heavy loads and/or where deflection control is critical. On the other hand, timber has a lower density than steel or concrete and on the basis of specific strength (strength/density), it compares more favourably.

Timber has a lower coefficient of thermal expansion. This is an advantage because timber decks can be designed as pin-jointed or as semi-stiff jointed structures, where any rotation or movement at the supports can be accommodated by deflection of the members. In contrast, steel and concrete bridges usually have to be designed with bearings and expansion joints to cope with the movements caused by temperature changes.

Timber possesses a thermal conductivity in the region of one tenth that of steel or concrete and is therefore less affected by temperature changes in the first place.

Among the factors responsible for the reduced use of timber in bridge building, is the availability of durable varieties. Originally the low demand for timber meant that the existing sources (many of them ancient woodland) could supply the required quantity and quality. With the industrial revolution and the growth of global trading, the range of available types of timber multiplied at the expense of large areas of forest. This level of exploitation is now recognised as being unsustainable and the majority of structural timber now comes from managed sources. However, a side-effect of the development of modern tree farming is that trees are often felled when the rapid growth period ends, before they have reached full maturity. This means that large section timber has become less common. In the nineteenth century, structural timber was commonly of pitch pine, Baltic pine or Douglas fir. Oak was the most commonly used hardwood. Timber piles were of extremely durable timbers such as greenheart or ironbark, nowadays some of the tropical timbers such as iroko and sapele are being used.

An example of problems with timber availability is that of the grade II listed Chinese Bridge, which crosses the River Ouse at Godmanchester in Cambridgeshire, shown in Fig. 8.6. When last maintained, it was not possible to obtain the required sections in Douglas fir, and consequently the more expensive European oak was used as replacement.

One important consideration regarding choice of a particular timber is its relative durability. It is a common misconception that hardwoods are more durable than softwoods *per se*. This is not necessarily the case, there are many hardwoods that are far less durable than some softwoods. Provided the structural and aesthetic requirements are addressed, it can be more economic to use preservative-treated softwood, in preference to durable hardwoods, such as greenheart.

Trusses

Design of the first timber truss is generally attributed to Palladio in the sixteenth century. Engineers developed different forms of timber truss in the nineteenth and early twentieth centuries, as shown in Fig. 8.7. In 1820, Ithiel Town developed the simple lattice truss, which was later used extensively in American covered bridges. The well-known truss systems developed separately by Howe and Pratt in the 1840s relied upon timber diagonal bracing with vertical iron ties and iron diagonal ties with vertical timber struts, respectively. The use of bolted iron rods enabled the provision of precompression to stiffen the joints between the main timber members, thereby improving the truss's performance.

The extensive use of timber trusses as beam elements in many bridges

Fig. 8.6 Chinese Bridge, Godmanchester, Cambridgeshire, 1827. Originally built in 1827, rebuilt to the same design in 1869 and 1960, the parapets have been renewed twice since the last rebuild, in 1979 and 1993

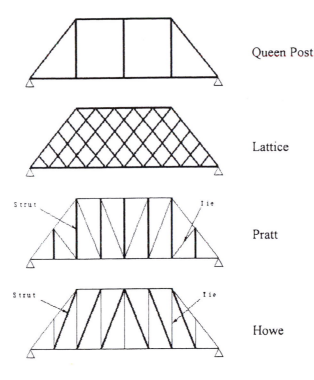

Fig. 8.7 Truss configurations

warrants some consideration of the evolutionary process by which the more recent, easier to maintain, better performing truss forms were developed.

The period when the greatest number of timber truss bridges were built was in the late nineteenth century, many being built in New South Wales, Australia (see Fraser, 1999). This was a direct result of the combination of the availability of timber and the low capital cost of timber bridges in comparison with other structural materials. The New South Wales Government issued a decree, stating that all road and rail structures, particularly bridges, were to be built from local timber.

The first widely used truss form was the Queen Post truss, also known as the Old PWD truss. In practice, there were a number of problems arising from the use of the these trusses which further stimulated the development of new truss forms, namely:

- low availability and difficulty in handling large, single-sized segmental arch components
- lack of redundancy in the top cord necessitating propping and closure of the bridge, prior to member replacement
- no account was taken of the higher stresses imparted to the vertical iron ties at the ends of the truss
- the inability to replace individual planks or 'flitches' that comprised the mechanically laminated lower cords
- drying shrinkage caused extensive sag or deflection.

Various designers sought to address these problems, the first being John McDonald who introduced a number of design improvements:

- the elimination of the use of a 'double thickness top cord' inner truss
- splaying the ends of members to provide stability
- the use of cast iron clamps (instead of drilled holes) to connect vertical ties which enabled a better joint to be constructed
- the use of cast iron end-plates on diagonal timbers which improved the force transfer at joints
- a system of steel wedges built into the design, so that shrinkage gaps could be closed
- development of a system for temporary tying and propping, thereby creating an alternative load path, facilitating removal of principal members without bridge closure
- the use of cheaper, more readily available sawn timber in place of hewn or cleft timber
- composite design which utilised more fully the different strengths and weaknesses of steel and timber.

The next significant design developments were made by Percy Allan. Design improvements employed in the Allan truss include:

- improved triangulation (a design approach which led to the development of the Howe truss)
- the use of smaller, shorter and more readily available sections
- the use of spacers between timbers (as opposed to laminated members) to improve drainage, increase the buckling strength and make painting easier
- improved, more economic provision and design of vertical tie clamps
- the use of cast iron shoes at all joints for good load transfer
- a means of tightening the truss through bolts on the vertical tie clamps
- the use of more members giving increased redundancy making member replacement easier

Following these improvements, the composite truss was further developed by De Burgh and Dare, incorporating steel lower cords. These improvements were significant, but were not sufficient to enable the bridges to meet demands that were to be placed on them in the future, such as:

- increase in traffic loading
- increase in traffic volume
- increase in traffic speed (this affects the fatigue and impact performance)
- decay of timber compression members from lack of drainage
- decreased availability of suitable timber
- increasing costs of labour and maintenance costs.

It is most likely a combination of these factors that contributed to the gradual replacement of the timber truss bridges in New South Wales with steel or concrete bridges.

PERFORMANCE

Physical defects of timber

Common physical defects that can affect the performance of structural timber are illustrated in Fig. 8.8. These are:

Check   Radial, longitudinal splitting cause by residual stresses from seasoning

Crack   Faults across the grain resulting either from external forces or unequal longitudinal shrinkage

Shake   Grain separation between rings of annual growth

Split   Resulting from global stresses on the member.

In addition to the above, the material properties of timber which can affect its performance, are volume changes due to moisture movement, its susceptibility to degradation and the fact that timber is anisotropic. One form of degradation that can affect exposed timbers is their susceptibility to deterioration under the action of the ultraviolet content in sunlight. This only affects the topmost surface, but can enable some forms of biological decay to take hold, and possibly increases the propensity of the timber to absorb surface moisture.

Moisture movement

Timber, when first cut, may contain in excess of 28% moisture by weight. Timber is hygroscopic and any change in moisture content will result in dimensional changes in the section, as well as affecting the strength and stiffness of the member. Furthermore, these changes will be directional. Prior to machining for structural use, timber should be conditioned to a point where the moisture content is in equilibrium with its intended environment. Some traditional forms of construction depart from this rule to make use of the shrinkage that accompanies drying. Within the wood, moisture is retained in the tracheids (in the case of a softwood) and can travel along these hollow cells faster than it can travel between them. Consequently, because structural members are always designed with the grain running along the member, changes in moisture content will have greater effect on the cross-sectional dimensions of the member than on its length.

It is not surprising therefore, that construction using 'green' timber (freshly cut) is liable to substantial distortion. Generally, the design should accommodate the instability of the timber through articulation and jointing

Fig. 8.8   Timber defects    **Check**  **Crack**  **Shake**   **Split**

techniques. One such case is illustrated in Fig. 8.9, where construction of an oak footbridge was commissioned with the brief that it was to be identical to a previously existing (but since removed) bridge in the grounds of a country park. Due to funding and planning constraints, there was insufficient time for the purchase and 'laying-up' to season sufficient timber, as this would have taken several years. Consequently, the bridge was constructed of green oak, with a certain amount of distortion anticipated. In the event the distortion duly took place.

Since fasteners and joints are most often fitted to the ends of members, and since end grain is most affected by changes in moisture content, the potential exists for the greatest moisture damage to affect the area subject to the highest local stress. It is important therefore in timber design to control exposure to moisture.

Creep

When under load timber exhibits creep, which is defined as time-dependent deflection. Creep is influenced by load and the moisture content of the timber. The nature of timber means that the moisture content is seldom uniform throughout an entire member. There will inevitably be a gradient of moisture content, with the outside and end sections fluctuating to a greater extent than those near the centre of the member. Creep is therefore difficult to predict with accuracy and can easily be underestimated. The design codes classify service conditions that depend upon the moisture content of the timber. Timbers in external situations almost certainly come under the more aggressive service classes, unless the structure is roofed and well-ventilated.

Fig. 8.9 Newly constructed oak footbridge at Standen quarry. (Note: wire mesh infill to parapets)

Table 8.1. Biological degradation

Type	Symptoms	Susceptible Conditions
Sap stain, non-rotting	Blue-black staining	Exposed softwoods
Serpula lacrymans, rotting (dry)	'Mushroom-like' smell, red dust fungal growth, distorted timber	Poor ventilation, high humidity
Coniophora puteana, rotting (wet)	Thin black-brown strands on surface, cracking of timber	Damp environments with high water retention

Biological degradation

Biological degradation is an inevitable consequence of the organic nature of timber. Both rotting and non-rotting fungi can affect timber (Table 8.1).

Although dry rot is a far more serious and rapidly spreading form of decay, and can pass through masonry, the action to be taken for any timber affected by either dry or wet rot is similar. The affected timber should be cut out, the cause of the rot removed, surrounding areas treated and new pre-treated timber, with a moisture content similar to that of the parent section, substituted.

Wet rot (see Fig. 8.10) can occur when the in-service moisture content is more than about 20%. The in-service moisture content can vary throughout sections and structures, dependent upon the service environment. Wet rot is by far the most likely form to be encountered in the timbers of a bridge.

Fig. 8.10 Substantial wet rot to 300 mm × 150 mm, pitch pine, railway sleepers

The need to control moisture is compounded by the support it offers to various agents of biological attack. Fungal decay from airborne spores is an ever present risk, and poor drainage and debris accumulation can encourage the build-up of spore-infected water which in turn will facilitate the growth of fungi, causing members to rot.

Staining alone poses no threat to structural integrity, however it indicates the existence of moisture, which can affect the structural performance of timber. An example of staining is shown in Fig. 8.1.

The custodians of the Mathematical Bridge, Cambridge, have found that the oak tends to attract green algae, and that this can effectively be removed by water jetting.

Depending on the type of timber, permanent submergence in a non-aggressive aqueous environment can be less onerous than cyclic wetting and drying. Timber is more likely to decay if it is alternately wet and dry. This environment tends to prevail at the ground level and splash zone, hence for assessment it is necessary to excavate to a depth of approximately 0.5 m around piles or posts and inspect the timber for decay. An example of treatment to protect timber in the vicinity of the splash zone it shown in Fig. 8.11.

Moisture can be retained in the splits and checks that can occur in timber as a result of dimensional instability. One possible means of addressing this threat is through the repeated application of a waterborne fungicide, insecticide or mould inhibitor, having first assessed the long-term costs in terms of labour and possible damage to the environment. Some water-based stains and pigmented coatings can provide this protection. Care is required in handling preservatives, for example chromated copper arsenate (CCA) solutions are toxic, up to the time when they become fixed in the timber.

Five hazard classes for wood and wood-based products are described in

Fig. 8.11 Preventative design measures. Protection of timber in the splash zone

the British and European codes, dependent upon the general service conditions (e.g. proximity to water). This enables the designer to establish the required design considerations for the various members that comprise the structure. For example, with any timber under the lowest hazard class, provision should be made for attack from beetles, possibly also termites (depending on location), but no other biological species.

Anisotropy

Due to its cellular structure, timber has anisotropic properties and in this respect is more akin to modern fibre composites than to steel or concrete. This can result in large differences in the elastic modulus and strength of the material when measured parallel and perpendicular to the grain. Performance of the various jointing techniques is also affected, though not necessarily in an identical fashion. It is therefore important that the direction of the grain is taken into account in the assessment and design of joints and members. As already discussed, the effects of moisture movement are also anisotropic.

Anisotropy is most evident in cleft timber, which can be much stronger than sawn timber. This is because cleaving requires a straight grain structure, and freedom from knots. It is the freedom from such imperfections which effectively raises the strength of the section. Cleaving was once fairly common, but is rare now due to high labour costs and shortage of suitable timber. It is worth noting that timber that cleaves easily will be difficult to nail or screw, since the nail or screw is likely to act as a wedge, causing the timber to split. The cleavage properties of timber are determined more by grain interlocking than density. Occasionally, the cleaved timber will have grain that is curved in the plane perpendicular to that in which it is cleaved, thus effectively creating a precambered section.

TIMBER JOINTS

Steel straps

Steel straps (as shown in Fig. 8.12) have long been used with structural timber, and can be relied upon to blend with the renovated structure as they weather. Steel plates attached to opposite faces of a timber can be stitch-bolted together to produce a neat attachment. Stitch bolting entails the application of compressive stress between two opposite recessed plates, held by tensioned bolts (see Fig. 8.13). In the design of these mechanical tensioned repairs, it is necessary to consider the bearing capacity of the timber, which could easily be crushed by over zealous tightening of the bolts. It is also worth noting that such joints can become loose due to moisture movement. This arises when the member absorbs moisture, but is prevented from expanding so that a compressive stress and permanent strain is developed. This stress will then relax upon drying out and as there is no restraint to prevent shrinkage strain, the member then becomes loose between the compressive plates.

Fig. 8.12   Mild steel
strap (as used on the
original structure)
joining a greenheart pile
to a reclaimed pitch
pine upper cord, at
Wickham Bishops
Viaduct

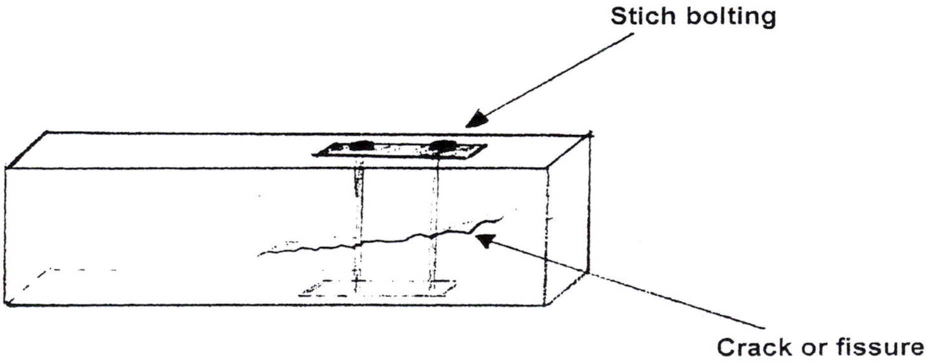

Stich bolting

Crack or fissure

Fig. 8.13   Stitch bolting

Adhesive joints

Adhesives are a fairly recent development being generally more suited to
well-machined softwoods and their use is not always feasible in older, traditional timber structures. Adhesives should only be applied to sound surfaces having no loose material and being free from grease and moisture.

Typical joints where adhesives are used include scarves, fingers, dowels and splices. Unsurprisingly the greater the bond area, the stronger the joint. For this reason the shallow scarf joint provides the highest strength, although being shallow, it also requires the greatest length of sound timber on which to bond. Scarf and finger joints are used in glulam construction to provide theoretically unlimited, continuous lengths of member.

Resin-grouted epoxy repairs are known to perform badly upon wetting, with a potential decrease in strength compared with the original, dry, solid timber member in excess of one third. Any structural adhesive to be exposed in all weathers should obviously be waterproof, and is most likely to be of the thermosetting resin type.

Design codes prescribe suitable adhesives for specific environments. The timber variety will determine to some extent the ease of adequate bond formation:

- oily timbers such as teak require special consideration
- preservative treatment of the timber can affect adhesive behaviour, some preservatives contain water repellents, which prevent the adhesive from wetting the surface (in the manufacture of glulam, such preservatives are not permitted)
- there are some glues which will not set in the presence of tannic acid, as found in oak and Douglas fir.

As with any adhesive application involving wood, it is necessary to strip or plane the outer surface prior to gluing, to ensure that a good bond can be achieved. Also the moisture content in the timber should be below 20% and close to the in-service level. Subsequent preservative treatment of a cured joint is unlikely to affect it. The satisfactory on-site use of adhesives for structural joints requires the following measures to be observed:

- maintain strict quality control and adherence to instructions provided by adhesive manufacturers
- meet the ventilation and other requirements of Health and Safety
- maintain adequate temperature and pressure for curing the adhesive. In smaller members, nailing may provide sufficient pressure, whereas larger members are more likely to require clamps, screws or bolts
- measure and monitor the moisture content of the timber
- provide a good, clean, dry surface, degreasing the timber if necessary with a non-water-soluble agent and neutralising acidic timbers with a 1% solution of sodium carbonate.

It is worth noting that a correctly executed adhesive joint is likely to have a fatigue performance significantly better than is attainable by mechanical methods. Also, it is extremely difficult to make a satisfactory adhesive joint outside the controlled environment of a factory.

Traditional joints

Some of the older types of mechanical joint, such as the pegged mortise-and-tenon, can provide an extremely rigid, attractive and durable joint. In any case it is preferable to use traditional methods for conservation of old timber bridges. The tenon needs to provide shear capacity and sufficient area to permit installation of adequately sized pegs. The tenon peg holes are often cut slightly closer to the member (i.e. not concentric with those in the mortise), in order to cause the joint to tighten upon installation of the pegs. If the timber has been used in its fresh-cut (green) state, the residual stresses set up from drying out can result in longitudinal shrinkage, so that the member tries to pull out from the mortise. The structural effect can be to tighten a framework, although it can also generate unequal strains, resulting in torsion and various undesirable forms of instability.

Another traditional form of joint is the mechanically fastened scarf joint (as used on the Wickham Bishops Viaduct). The joint is held together by compressive stress from bolts (see Fig. 8.14).

Metal fasteners

Due to the required level of craftsmanship and expertise (both design and workmanship), the traditional methods of jointing are not always practical for large volumes of work. Recourse is therefore necessary to contemporary metal fasteners such as nails, screws and hangers, shear plates and toothed plates. As with adhesive joints, care must be taken with regard to the in-service moisture content, load duration, timber varieties and so on. Toothed plate connectors are invariably of galvanised steel, and almost exclusively used on softwoods, in timber-frame construction. There may be possible applications in footbridges, but metal fasteners are unlikely to be permitted for conservation work on historic bridges.

One timber-specific problem that can affect both mechanical and glued

Fig. 8.14 Wickham Bishops Viaduct. Scarf repairs in reclaimed pitch pine to 300 mm × 150 mm, bracing members

joints occurs when two deep sections are joined at 90°. The surface of the timbers at the joint is restrained from movement (i.e. expansion due to moisture uptake). The anisotropic performance of the timbers would result in greater strain transversely to the grain (compared with that in the longitudinal direction). Where this strain is prevented by restraint from the joint, large stresses are created in the timber as it tries to expand. It is this stress that can result in the longitudinal splitting (with the grain) leading up to the joint.

Methods of jointing are summarised in Table 8.2.

CONSERVATION

Replacement of members

The factors to be considered when assessing whether to replace a timber member fall into three broad categories, as summarised in Table 8.3.

The replacement of members on a like-for-like basis is a preferred method of conservation as the appearance of the structure is unchanged, and the original design and type of material are retained. However, it is not always feasible and compromises may have to be made.

It is necessary to assess the structure to determine the condition of the members and extent of deterioration. There may be a small number of members requiring replacement, or the deterioration may have reached the stage

Table 8.2.   Methods of jointing

Method	Advantages	Disadvantages
Clamp and stitch bolt	Weathering can lessen changes in appearance. The appearance is neat and complementary.	The joints can loosen due to moisture movement. Changes the appearance. Loss of originality.
Splice joints	Can be designed to have a good appearance.	Labour intensive.
Adhesive joints	Good joints can be made under factory conditions.	Not recommended for *in situ* application to old and weathered timber.
Traditional joints	The most suitable joints for historic structures.	Expensive. Require good craftsmanship.
Resin-grouted repair	A quick cheap fix. Little change to appearance of the timber.	Adhesion to the timber can fail under damp conditions.
Contemporary mechanical connectors	Little change in appearance.	Loss of originality.

Table 8.3. Replacement of members

Task	Considerations
Selection of timber	Maintenance records should be studied to determine the original variety of timber as it may have been replaced at a later time. It may not be possible to obtain the required variety of timber in the required sizes. The timber should be obtained from a sustainable source. If possible, members removed from the structure should be cut down and re-used.
Ensure durability	Durability performance can be improved without spoiling the structure: • timber can be obtained having factory-impregnated preservative treatment. • design detailing can be improved to avoid traps where detritus can collect. • water management can be improved.
Repair operations	The structure should be assessed to determine whether it can be carried out without closure to pedestrians and traffic. If not, it may be necessary to install temporary strengthening or propping. Constraints such as acceptability of lane closures for highway bridges, and track possessions for rail bridges may dictate the repair operations.

where there is little that can be retained and reconstruction is necessary. If it is established that replacement is the preferred choice and the structure has sufficient merit to justify what may be an expensive option, it will be necessary to determine the feasibility of replacements.

The service history of the structure should be studied to check whether the original variety of timber is in place. It is not uncommon for a different variety to have been used for later repairs and this may be a cheaper substitute that is inferior to the original timber.

It may not be possible to obtain the preferred variety, in which case a suitable alternative will have to be selected. Also, there may be difficulties in obtaining timber in the sizes required. The timber should be obtained from sustainable sources having Forest Stewardship Council approval. When the members being replaced are of different sizes, it may be practical to cut down and re-use some of the larger ones.

It will be necessary to carry out structural analyses to determine whether the replacements can be carried out without having to close the bridge for long periods or install temporary supports.

In designing the conservation scheme it is appropriate to seek to improve the structure. If it is required to raise the load-carrying capacity, strengthening measures will have to be taken. With the increased awareness of durability problems, it will normally be possible to increase the service life without spoiling the originality or appearance of the structure. Timber having factory-impregnated preservative treatment can be used and improved design details can be introduced. Design improvements mainly concern improved water management and avoidance of traps where detritus can collect.

There may be occasions when it is required to retain members, irrespective of deterioration. In that case sister members can be fitted as described in a later section in this chapter. In a different context, deteriorated members can be left in place and strengthened with a concrete skin as for Barmouth Rail Viaduct, see Fig. 8.18.

In cases where the structure has less merit it may be acceptable to substitute steel, for example, in place of large-section timber in main longitudinal beams.

Preservative treatment

The British and European codes classify over one hundred different timber varieties into three durability ratings:

- resistance of the heartwood to wood destroying fungi
- resistance to common insects and
- resistance to attack from termites and marine borers.

The required minimum natural durability classification dependent upon the hazard class of the service environment is given, as are the requirements for preservative treatment. In the codes a method is provided, by which the extent of penetration and retention of preservatives may be measured. To this end, there are nine classes for penetration requirement and various levels of preservative retention of between 10 and 25%. Provided the supplier of the preservative adheres to the code in the classification and labelling of the product, it will be possible to accurately specify an adequate preservative for the job. Protection from marine borers can sometimes be achieved with CCA, but certain species of borer are unaffected by this treatment. It is therefore important to correctly identify the species in cases of infestation.

When considering the replacement of timber members, it is necessary to go through the design process outlined in Fig. 8.15, to ensure suitability of the specification. One other main concern regarding the selection of preservatives is their effect on connections and fasteners. There are preservatives such as CCA-containing salts, which can cause rapid corrosion of ferrous connectors. This is more serious when the preservative has not been 'fixed' into the timber. One possible though costly measure is to use stainless steel connectors.

Various commercially developed, and in some cases proprietary, factory methods exist for the treatment of timber, for example through diffusion treatment or pressure methods such as the empty cell, full cell and double vacuum methods. These methods can accurately and predictably achieve the required extent of treatment and are classified as penetrative.

The various methods of *in situ* application, such as painting or spraying,

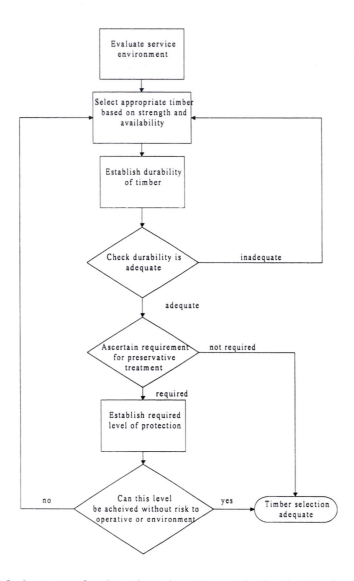

Fig. 8.15 Preservation
design flow-chart

are classified as superficial as they do not provide the depth of penetration
available through factory processes. The success of *in situ* painted or
sprayed preservative treatment is hard to quantify, and when there is a
choice, timber treated by factory methods should be used in preference.

In some instances, it can be beneficial to drill a grid of small holes into
the decayed timber to ensure adequate penetration of the preservatives into
the affected area. A commercial method involves inserting a boron 'poultice'
which then slowly permeates the timber. A major concern pertaining to the
in situ application of preservative treatments, and water-based weathering
coats, is the likelihood that they may cause the timber to swell and distort.

It should be noted that the health and contamination risks associated with
the *in situ* application of preservatives are far greater than those associated
with factory processes. Environmental legislation places restrictions on the
use of various preservatives, for example, the organotin compound tributyl-

tin napthenate is a prescribed substance under the Environmental Protection Act 1990 (EPA), consequently any potential users are obliged to obtain authorisation. Particular contamination risks exist in using these materials over rivers and in built-up areas. The Best Practicable Environmental Option, as expressed in the EPA, is site-specific. Therefore, it is not possible to provide blanket 'best-guidance' on such issues.

The use of creosote is no longer environmentally acceptable in aquatic environments.

Painting

Painting has a dual role of giving an attractive appearance in keeping with the intentions of the designer, and providing protection to the timber.

As with the application of preservatives, it is imperative to ensure that any paint (or stain) that is to be applied is suitable for the variety of timber and the local environment. The various types of paint that are available tend generally to fall into two classes: solvent-based and water-based. Technology in this area has developed greatly in recent years and the paints and stains of today can offer better performance than many older products. One prime concern when painting or staining timber must always be to avoid trapping moisture. The commercially available water-based paints and stains tend to be breathable and so in this respect have an advantage over some solvent-based products.

The replacement of the parapets on the Chinese Footbridge at Godmanchester (Fig. 8.6) is an example, where the requirement was for white railings. The traditional approach would have been to specify exterior gloss paint. Products are now available that contain pigment and so are opaque like a paint, yet are bound in a breathable, (sometimes) thinner matrix, and are less susceptible to failure through cracking. One of these products was used in this instance. The timbers, which were of iroko were first degreased with cellulose thinners. One coat of a solvent-based primer was applied (to prevent transport of staining leachate from the timber), followed by two coats of aqueous based pigmented stain. Iroko is eminently suitable for exterior joinery and would have been adequately durable without any treatment or finishes, but the requirement was for white parapets to preserve the original style and appearance. Wherever possible, stainless steel was used for the connections, due to the use of oak for the other members. The reason for this is that oak sap contains high levels of tannic acid which would attack unprotected steel.

Detailing for durability

There are many opportunities during the design process to incorporate details to achieve improved durability. These are summarised in Table 8.4.

An example of a durable detail, also used on the Chinese Bridge at Godmanchester, is the provision of a waterproof membrane on the tops of all support members in contact with the deck planks. The need for this arose because of the large number of dogs relieving themselves on the posts which supported the handrails.

Some examples of bad practice in the application of remedial methods for bridge conservation are given below.

- Joint repair using thixotropic epoxy resin and steel dowels or plates. This takes no account of the hygroscopic behaviour of timber resulting in possible breakdown of composite action following shrinkage of the timber.
- Plugging of fissures, thereby restraining the timber and imparting more severe stresses in the member, should it subsequently close up.
- Reckless use of preservatives, causing damage to the environment.
- Lack of an adequate, regular maintenance and inspection plan; for example removal of debris and clearance of drains.

Table 8.4.   Methods of achieving improved durability

Method	Aim	Action
Enclosure	Minimise exposure to weathering.	Provide roof or cover. Apply protective concrete skin as in the case of Barmouth Viaduct. Although in this case the method was almost certainly the best choice, it is a very special case, and should not be regarded as having general applicability. Any repeated use of such a method would require careful re-evaluation prior to its specification.
Material selection	Ensure adequacy for purpose.	Check inherent durability of timber variety, preservative requirement and moisture content at installation.
Details	Avoid distortion due to water intake.	Protect horizontal surfaces, joints and end-grain, use flashing. Detail for rapid drainage, avoiding ponding and saturation of timber. Use waterproof membrane.
Preservatives	Prevent biological degradation.	Adopt factory applied, penetrative treatment systems wherever possible, use *in situ* methods with care.
Surface treatment	Prevent weathering and reduce fissuring of surface by shielding the timber from potentially harmful UV rays in sunlight, and encouraging rapid water run off.	Apply pigmented maintenance coats, most likely to be in the form of an aqueous-based stain, since impermeable barriers can seal moisture in and are likely to cause problems.

The work carried out on the historic covered bridges in the USA provides a good example of a successful programme of conservation. The provision of a roof over the structure arose from concern regarding durability. This approach was extremely popular in North America in the nineteenth century. Timothy Palmer's 1806 'Permanent Bridge' over the Schuylkill River, USA was the pioneer of this design. There were almost 400 covered bridges in the state of Kentucky in the mid-nineteenth century, but only thirteen remain. In recognition of their historical importance, a 3-year, $1.5 m inspection and restoration project was carried out between 1996 and 1997. As with any work on a structure of cultural or historic interest, it was important to establish the original design criteria and full details of the maintenance history. Conservation of the covered bridges was no exception and was guided by information obtained from exhaustive desk studies and inspections. In addition to checking the roof and walls to ensure that the bridge was still protected from the elements, the structural members and connections were also inspected. Presence of timber decay was identified through visual inspections and impact-echo testing (a dull thud indicating the possibility of decay). The extent of decay was determined through drilling (see section on inspection techniques: microdrilling). The connections were checked for tightness, corrosion and mechanical wear. The original timbers used in many of the bridges were 'mature' yellow poplar, which was no longer available in the required size and quality. In this instance, Douglas fir was used as a replacement timber, as its properties are similar. For sections susceptible to pooling of water, insect attack and decay, such as the deck, deck joints and connections, American white oak was used due to its high resistance to insect attack. One over-riding concern regarding the supply of timber to these contracts was the contractual requirement for timber with a moisture content of 15% or less since the shrinkage resulting from drying out (obviously more of an issue with covered bridges) can cause serious structural problems within months of construction. A moisture content of 15% is low for any external timber exposed to weather and probably reflects the shelter to the superstructure afforded by the roof.

'Sister' members

Various methods exist to connect and repair joints and members. Of those applicable to *in situ* repair and strengthening, there is substantial variation in the resulting appearance. The most visually damaging and least acceptable should only be used for temporary or emergency repairs, and consideration given to any potential long-term disfiguration that may result. Such methods include the addition of a 'sister member', see Fig. 8.16, clamping, exterior reinforcement, splicing and propping. 'Sister member' is a term used for a member installed in parallel to a deteriorated member, so that the load is either shared or transferred altogether to the new member.

'Sister members' are sometimes used as temporary measures to be re-engineered later. To that extent they can be regarded as props. There could also be occasions when the original members are of sufficient heritage importance that they must be retained and replacement is not an option. In that case installation of 'sister members' would be a conservation method to be considered.

Fig. 8.16 Decayed pile column. The left-hand column was strengthened with two sister members

Protective sleeves

Protective sleeves, or encasement, can be used to protect and strengthen deteriorated members as, for example in the conservation of the Barmouth Rail Viaduct. This can be accomplished with either metal or concrete encasement of the member under threat, thereby protecting it from agents of attack and if required, providing an increase in structural section. Clearly this method of renovation must encase the entire affected area plus a safety region, and as mentioned, is of limited application.

Post-tensioning

In the case of longitudinally laminated wooden decks, both deflection and the passage of moisture can be reduced by the installation of post-tensioning. The need to protect both the timber and the post-tensioning material from moisture and corrosive agents should always be addressed. Similarly, the requirement with Douglas fir and oak to use non-ferrous fasteners, because of the extreme corrosive effect of the high tannic acid levels in these timbers, should also be addressed.

Post-tensioning can be used in its better known role to reduce tensile stresses and strengthen the bridge. When applying the compressive stress, care must be taken to ensure that the timber member is not overstressed and damaged.

In general, post-tensioning changes the structural action of the bridge and can affect its appearance. Such changes are not readily agreed by heritage authorities. Methods of repair are summarised in Table 8.5.

Table 8.5. Methods of repair

Method	Advantages	Disadvantages
Sister member	Permits original timber to be retained.	Changes the appearance.
Propping	A good temporary measure to improve safety.	Must be replaced by a permanent scheme.
Protective sleeves	Repairs, strengthens and protects deteriorated members.	Changes the appearance. Loss of originality.
Post-tensioning	An easy method of making a quick repair. Reduces deflections and passage of moisture.	The tie-rods or tendons require good protection against corrosion. Must be carefully designed and applied to avoid overcompressing the timber. Loss of originality.

Examples of conservation

Wickham Bishops Rail Viaduct, Maldon, is a beam and trestle structure built across the River Blackwater in 1848, see Fig. 8.17.

The railway was closed in 1966 and the viaduct left to decay under the actions of weathering and vegetation. The worst deterioration was at water level, where the timber piles lost a substantial amount of cross-section. The required load carrying capacity was vastly reduced when the viaduct was no longer required to carry railway vehicles, rendering half of the pile columns redundant. In 1995, when the viaduct was repaired, the redundant members were left alone and other pile columns were replaced like-for-like. Bitumen paint was applied to protect the timber in the vicinity of the splash zone.

Fig. 8.17 Wickham Bishops Rail Viaduct, Maldon, Essex. Built in 1848, closed to traffic in 1966 and refurbished in 1995. A scheduled ancient monument.

Barmouth Rail Viaduct, Wales, was constructed in 1866, see Fig. 8.18. The structure was temporarily closed when it was found that the timber pile columns had reduced cross-sections as a result of borer infestation. The deterioration was at ground level where pools of trapped water encouraged the infestation. In 1986, when the viaduct was repaired, the deteriorated lengths were strengthened and protected by addition of 3 m long glass-reinforced concrete jackets. Cementitious resin was grouted between the jackets and piles, killing off the Teredo worms, and strengthening 330 of the 500 piles. Where this treatment was not feasible, forty-eight of the piles were replaced like-for-like with greenheart, solid timber piles. A continuing programme of staged replacements and repairs to the superstructure was being carried out in 2000.

Angle Vale Bridge, Australia is a mechanically laminated arch structure built in 1876. It was found to be understrength and traffic was diverted, but the deck was subsequently found to have insufficient capacity for pedestrians due to rot, splitting and the action of termites. In 1987 when the bridge was repaired, members having over 50% loss of section were replaced. Minor deterioration was repaired with epoxy grout. Water-based diffusion preservative was applied to the existing timber prior to application of two sprayed coats of copper napthenate oil to protect against the weather. Durable (and expensive) timber was used for the decking to achieve an anticipated extension in life of between 50 and 100 years. Sound bolts were cleaned, rethreaded, oiled and re-used.

Selby Swing Bridge is a movable structure across the River Ouse, originally built in 1791 and reconstructed in 1970, see also Chapter 11. In the reconstruction each pile bent was formed from eight hewn greenheart piles, 406 mm × 406 mm in cross-section and 21 m long. The piles were cross-braced with opepe timbers and topped by metal connectors to steel I-beam

Fig. 8.18 Barmouth Rail Viaduct, Wales, 1866. Substantial repairs were carried out in 1986

crossheads. Timber joints were made using galvanised 25 mm or 32 mm diameter bolts with 102 mm square washers. This form of connection was chosen in preference to smaller bolts as it was considered that larger bolts would offer more effective resistance to long-term corrosion. In addition, the members would be easier to erect as there would be less cutting of the mating surfaces and the fit would be less critical. The fendering was independent of the main structure and composed of pitch pine piles. They were braced diagonally and horizontally with pitch pine timbers. Fender walings were 259 mm × 127 mm spaced at 1.22 m centre-to-centre vertically. The timber deck was of 38 mm Douglas fir laminates preserved by immersion in a creosote bath for 2 min. The road wearing surface was 19 mm Finnish birch grade WBP plywood faced with calcined bauxite chippings set in a matrix of epoxy resin. The footpath surface is untreated kéruing timbers with opepe hardwood kerbing.

MANAGEMENT PRACTICE

Inspection

The nature of biological decay can make a meaningful inspection difficult. For example, it will not always be easy to ascertain the extent of interior damage to a member resulting from an infestation of boring insects. In these circumstances, depending on the experience of the inspector, it may be prudent to seek the advice of a wood technologist. Methods of inspection are summarised in Table 8.6.

The numerous unknowns (resulting from natural variations in the material) mean that more conventional non-destructive techniques, such as impact-echo, endoscopy, thermography, X-rays, gamma rays and ultrasonics are not always a viable means by which to investigate the internal condition of timber. This was borne out by the case of Barmouth Rail Viaduct (see Fig. 8.18). In this instance, ultrasonic tests failed to differentiate wormholes from timber grain and could not be used underwater. X-rays suffered from a lack of definition, and low frequency sonic tests failed to give the required correlation. In the event, the only means of inspection found to be reliable were intrusive. Sections were either cut open or small inspection holes were drilled.

In contrast, to the negative experiences at Barmouth, several nondestructive methods were successfully used on the Angle Vale Bridge, in Australia. Ultrasonic testing was used to determine the extent of timber deterioration, radiographic testing of the bolts was used to establish corrosion levels and static load testing was carried out to assess the stress–strain behaviour of the structure.

A nondestructive method that can provide very accurate information on the timber is the microprobe or digital microprobe. These are miniature drills capable of detecting decay, cavities, tree growth and the history of the timber. They rely on the accurate measurement of the torque required to penetrate the timber, and were used to good effect in the remedial work to Windsor Castle, following the fire in 1992. Although this equipment can be purchased over the counter, the skill level required to use it correctly requires experienced and knowledgeable operatives.

Table 8.6. Methods of inspection

Method	Advantages	Disadvantages
Nondestructive techniques	Can be carried out, without, causing any significant change in appearance. The different types of test provide information on different aspects of the state of the timber.	Do not always work satisfactorily. Results are sometimes confusing and require experience and knowledge to be interpreted correctly.
Invasive inspection	A direct and unambiguous method. Quick, accurate and cheap.	Provides local information which may not be representative of the overall condition of the timber.
Grading	Can be carried out visually and cheaply.	Can only be carried out on accessible components.
Identification of decay (visual)	Quick and cheap.	Only applicable to exposed external surface. Does not provide information about timber beneath the surface. Requires experienced inspector.
Moisture content (commercial equipment)	Quick and accurate.	Requires experienced and knowledgeable operative.
Load testing (assessment)	Provides information about structural behaviour. May identify hidden strength.	Expensive to carry out. Care must be taken to avoid causing damage. Results can be difficult to interpret.

Grading (a part of which involves the statistical estimation of the percentage of knots in a section) of new timber has been developed and mechanised, becoming more of an exact science, thus enabling more reliance to be placed on the given range of strengths. However, not all *in situ* members will be sufficiently exposed, even for reasonable visual inspection or grading.

Unquantifiable damage due to decay might also exist in the member, see for example Fig. 8.19. Additionally, large, old beams will invariably include a small central section of nonstructural pith. Identification of the timber variety (which can in practice be difficult), along with the effective cross-section, will enable the design strength to be derived.

The appearance of timber can easily be misleading to the untrained eye and there are occasions when the true extent of biological degradation can

Table 8.7. Some examples of surviving timber bridges

Bridge	Date of construction	Comments
Angle Vale, Gawler, Australia	1876	The last surviving mechanically laminated timber arch bridge in Australia. Repaired in 1940, converted to pedestrian use in 1960, later closed and fell into disrepair. In response to public pressure conservation work was carried out in 1988 and the structure was given an estimated life of 50 to 100 years.
Barmouth Viaduct, Mawddach Estuary, Wales	1866	Rebuilt in 1906 and substantially renovated in 1986. Typically, for railway viaducts of this time, it is of beam and trestle form, see Fig. 8.18.
Mathematical Bridge, Cambridge	1750	Twice replaced by replica structures, most recently in 1904. Although having the appearance of an arch, this bridge is structurally a truss, see Fig. 8.4.
Old Shoreham Bridge, West Sussex	1781	Beam and trestle originally built of oak. Rebuilt in 1916 of blue gum to a design similar to the original, see Fig. 8.3.
Rhinebridge, Shaffhausen	1755	An ornate bridge having an innovative form of truss-arch design.
Whitney Toll Bridge, Hay-on-Wye	1797	A traditional structure, having a three-span beam and trestle superstructure, between large masonry abutments. Replacement members comprise ekki (superstructure) and keruing (decking), both tropical hardwoods.
Wickham Bishops Rail Viaduct, Maldon, Essex	1848	Taken out of service in 1966. A scheduled ancient monument, see Fig. 8.17. Of typical beam and trestle form. Conservation work carried out in 1995.
Wooden Bridge, Broomhill, Inverness-shire	1894	A fifteen-span, beam and trestle bridge.

only be accurately established by a wood technologist. Age alone does not necessarily have any bearing on the load-carrying capacity of existing timber. When assessing deflection it is important to note the extent to which this is due to creep, as opposed to elastic deformation.

The moisture content of timber can be assessed using calibrated moisture meters comprising two electrode prongs either held to the surface, or ham-

Fig. 8.19 Example of typical decay. The decay has been from the inside, causing significant reduction in section

Fig. 8.20 Fort Augustus. A heritage-listed timber bridge in an attractive location

mered into the timber. Surface readings reflect a more transient condition than those at depth. The electrical resistance of the timber is dependent on the moisture content, the level being given as a percentage. Due to the hygroscopic nature of timber, coupled with the structural and biological effects of changes in moisture content, an accurate measurement at critical locations is imperative. Design codes assume a nominal moisture content, with the requirement to make adjustments to accommodate higher readings (and inspect those areas closely for signs of biological degradation). Moisture content will also be affected by changes in the ambient temperature and this is the only likely structural effect of such changes. In temperate climates, like that of Britain, the effect on moisture content of seasonal temperature changes is fairly small and may be insignificant. Checks are always necessary to ensure the continued provision of adequate ventilation to all members and joints.

One reasonably well-accepted means of taking inspection further is through load testing. Static loads are applied and the resultant deflections measured. These are then compared with calculated values and the differences will then permit estimation of, for example, the effective cross-section. Care must be exercised, however, when attempting to predict ultimate strength from the result of a load test as the behaviour of timber as it approaches its ultimate strength, particularly in compression, is highly non-linear. When carrying out load tests, the ICE Guidelines on Supplementary Load Testing of Bridges should be followed (see Chapter 12).

Maintenance

Water management, preservatives and structural repair have been discussed earlier in the chapter and are summarised in the following section.

Water management

- Provide good ventilation and drainage.
- Ensure that end grains are protected from moisture and weathering.
- Design details to permit good 'run-off' and having no moisture traps (for example, through poor fit or wear and tear).
- Maintain a maintenance schedule to prevent blocked drains and vegetation from creating moisture traps.
- Specify suitable timber treatment or protection to be used in places where long-term contact with water is anticipated.
- Make design allowances for movement and distortion due to moisture.

Preservatives

- Fully evaluate the service environment.
- Select appropriate timber and preservative treatment.
- Monitor treatment, i.e. check that required level of penetration has consistently been achieved.
- Adopt realistic approach to the in situ application of any treatment process.

Structural repair

- Conduct full desk studies to establish such relevant information as the basis of original design, load history, freedom to alter structure (whether heritage listed), environmental hazards (meteorological and infestations, etc.), environmental restraints.

- Establish scope for structural alterations in relation to heritage requirements.
- Discount schemes involving epoxy grouting, or for that matter, any form of site-applied adhesive.
- Aim to keep the repair 'honest' without being unsightly. This is most likely to be acceptable to heritage authorities.
- When there is a choice, use traditional methods in preference to others.
- Consider the relative movements of different materials in joint design.
- Consider the highly corrosive nature of certain timbers towards ferrous fasteners.
- Be aware of the hazard to health posed by work using some preservatives.

BIBLIOGRAPHY

Fraser, D. *Evolution of Timber Truss Road Bridges in New South Wales, Australia.* Proceedings International Conference on Historic Bridges to celebrate the 150th Anniversary of Wheeling Suspension Bridge, 21–23 October 1999, West Virginia University Press.

Chapter 9

Iron and steel bridges

<table>
<tr><td colspan="2">GLOSSARY</td></tr>
<tr><td>Iron ore</td><td>A naturally occurring mineral containing iron oxide mixed with sand, earth, clay and stones which contain silica and other compounds.</td></tr>
<tr><td>Smelting</td><td>The operation which changes iron ore into pig iron by the use of heat and chemical energy.</td></tr>
<tr><td>Pig iron</td><td>Crude iron as produced from a blast furnace and containing carbon, silicon and other impurities.</td></tr>
<tr><td>Blast furnace</td><td>A furnace used for smelting iron ore by means of heating a mixture of iron ore and coke or coal (formerly charcoal). A blast of air is driven into the furnace to enable the fuel to burn and produce a high temperature.</td></tr>
<tr><td>Cast iron</td><td>An alloy of iron, carbon and silicon produced by re-melting pig iron. The molten metal is poured into moulds to form castings of a particular shape.</td></tr>
<tr><td>SG iron</td><td>Spheroidal graphite cast iron (also called ductile iron). A cast iron developed since 1948 with improved properties which can compete with steel castings and forgings. Rounded graphite nodules are formed when the metal solidifies.</td></tr>
<tr><td>Wrought iron</td><td>Almost pure iron made by reheating and hammering pig iron to remove impurities.</td></tr>
<tr><td>Puddling</td><td>The process of making wrought iron carried out in a reverberatory furnace in which pig iron is melted and refined by the addition of iron oxide and other substances.</td></tr>
</table>

Reverberatory furnace	A shallow furnace used to melt pig iron, with the iron being kept from direct contact with the fuel. Used in the puddling process (for wrought iron) and open hearth process (for steel).
Steel	An alloy of iron and carbon produced from molten pig iron. Early processes used the Bessemer converter or open hearth, now replaced by the oxygen and electric processes.
Slag	The dross separated from the molten pig iron in a blast furnace or steel-making furnace.
Bessemer converter	A huge cylindrical vessel used in early steel making, into which molten pig iron was poured and air was blown through a number of holes in the base of the converter and forced through the molten metal. No external fuel was applied; the oxidation of the impurities and some of the metallic iron furnished the heat requirement. Certain additions were made when the molten metal was poured from the converter to assist the deoxidising of the steel and then anthracite was added to bring it to the correct carbon content. Used until 1970s.
Open hearth furnace	A large reverberatory furnace used in early steel making, where molten pig iron was mixed with scrap steel and iron oxide. A preheated gaseous fuel was injected into the furnace and similar chemical reactions took place as for the Bessemer process. Used until 1970s.
HSFG bolts	High-strength friction-grip bolts. Bolts designed to resist shear by developing friction between the plies and may also resist external tension. Do not require a lock nut.

BACKGROUND

Iron has been used in bridge construction since the turn of the eighteenth century and there are many fine examples in everyday use. When dealing with the conservation of these bridges, it is important to have knowledge of iron in its various forms, its manufacture and properties.

Cast iron

Originally, iron was extracted from ore in charcoal furnaces in a pasty condition in which it could be shaped and purified by hammering. With the invention of the blast furnace circa 1500 AD, it became possible to achieve the temperature of 1,250°C at which pig iron becomes truly molten and can be cast. It was not until Abraham Darby perfected coke smelting of local iron ore in Coalbrookdale in 1709 that it became practicable to produce reliable cast iron in large quantities. Others were slow to follow this example and, taking advantage of the local expertise and availability of cast iron, thirteen businessmen in the area proposed to construct a bridge over the River Severn to replace the overworked ferry. In 1776 a Bill was laid before Parliament for the construction of a bridge in cast iron. Thus was born the Iron Bridge, the oldest surviving bridge in the world to be constructed wholly of cast iron, see Fig. 9.1 (there were two earlier cast iron bridges in the north of England, but no longer extant). The bridge was a single arch of 30 m span, made from ten half-ribs, each cast in one piece by Abraham Darby III. It is suggested that the designs for the ribs and members of the bridge were prepared by Thomas Gregory, foreman pattern maker at Coalbrookdale and therefore a worker in wood, and this probably explains the construction being in timber style with mortise and tenon and dovetail joints. Construction of the bridge foundations started in 1778 and the bridge was opened on 1 January 1781.

The molten iron produced by the blast furnace was sometimes poured directly into moulds of the required shape, but this produced inferior quality castings. More often the molten iron was poured into a trough called a 'sow', which has numerous offshoots for casting blocks of iron called 'pigs', hence the curious term 'pig iron'. The same blast furnace could produce several varieties of pig iron according to the proportion of fuel used for smelting. There were six varieties of pig iron known by the following numbers and names and listed in order of decreasing carbon content and production cost: Nos 1, 2 and 3 grey iron, No. 4 or 'bright' grey iron (used for the manufacture of wrought iron), mottled iron and white iron. The process of remelting the pig iron to form cast iron components also permitted the iron to be refined by removing more of the impurities and it was also customary to mix different varieties of pig iron or pig irons from different

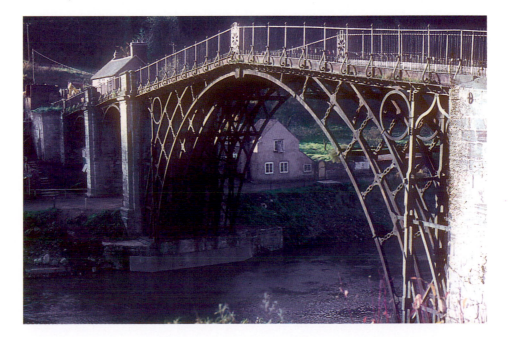

Fig. 9.1 Iron Bridge, Shropshire. The oldest surviving bridge in the world to be constructed wholly of cast iron, 1781

localities to achieve what was considered to be the optimum mix for a particular use. Virtually all the cast iron used for structural applications was grey cast iron because it was made from the best quality pig iron and was therefore most reliable.

Cast iron was seen by most engineers as the great new wonder material likely to supersede timber for all purposes because of its greater security in supporting heavy loads and in resisting fire and rot. Fire was a particular hazard at this time in the textile mills and Charles Bage designed a flax mill on the outskirts of Shrewsbury which was the world's first multistorey building with an internal cast iron frame (completed in 1797). The floor construction comprised cast iron beams with brick jack arches, which was to be later adopted for many short span railway bridges. Bage was a Shrewsbury wine merchant, amateur engineer and friend of Thomas Telford and his formulae were the first recorded design methods for cast iron beams and columns circa 1800.

A significant development in the late 1700s was James Watt's invention of the steam engine and the use of steam to power a strong blast of cold air into the blast furnaces. By the 1830s, a hot air blast enabled coal to be charged as fuel direct into the furnaces without prior conversion to coke and more iron could be extracted from old blast furnace slag. However, both practices introduced impurities and the quality of the cast iron was inferior. By making improvements to the manufacturing process over the next 40 years, the difference in quality between cast iron made by either a cold blast or hot blast became insignificant.

Cast iron contains typically 2 to 5% carbon, absorbed from the coke during smelting which makes it hard, brittle and very resistant to corrosion. Other impurities such as silicon, manganese, sulphur, phosphorus, chromium and copper also influence the characteristics of the material. The formation of grey iron is promoted by the presence of silicon and phosphorus and by a relatively slow cooling from the molten state. This allows some of the carbon to solidify and form graphite flakes surrounded by what approximates to steel (see Angus, 1976; Doran, 1992).

The stress–strain curve for a grey cast iron varies considerably according to the quantity and coarseness of the graphite flakes. The material under tensile stress has been likened to a slotted steel plate. The low elastic limit coincides with first localised yielding at the ends of the slots or internal notches. The graphite flakes act as planes of weakness that account for the relatively low tensile strength. Behaviour in compression is nearly elastic, with the graphite able to transfer compressive stress. Tensile and flexural strengths also vary with cross-section shape and size.

Thomas Tredgold published his work on the design of cast iron beams, columns and tie rods in 1822 based on his experiments. Unfortunately, he came to the conclusion that cast iron should be designed with the same maximum stress in tension and compression and the maximum tensile stress was approximately three times too high. Tredgold therefore advocated cast iron I-beams with equal top and bottom flanges and such beams were used in bridges up to the 1840s. Fortunately, in many cases, the beams were proof-loaded to stress levels sometimes very near their ultimate strength, but this did not matter provided that they passed the test and the subsequent loading was less onerous. The man who partly corrected

Tredgold's misconceptions was Eaton Hodgkinson who showed that the grey cast iron of his time was about six times as strong in compression as in tension and thus he evolved his 'ideal section' of beam with a bottom flange up to six times the area of the top flange. In practice, this proportion was generally found to be impractical and in most 'Hodgkinson beams' the proportion was nearer to 3:1 or 4:1. Hodkingon's beams were widely used from the 1830s onwards and his simple beam design formula was extensively used throughout the nineteenth century, however, it did overestimate the strength of larger beams.

During the years from about 1830 to 1850, many combinations of cast iron and wrought iron were tried out to overcome the shortcomings of cast iron. The use of cast iron beams declined during the period 1850 to 1890 due to the adverse publicity following the collapse of the Dee Railway Bridge in 1847 and the advances made in the use of wrought iron. Cast iron columns continued to be made for limited use up to the early 1930s.

Wrought iron

By about 1750, Abraham Darby II was making small quantities of wrought iron from coke smelted pig iron, which could be used in forges. The process of reheating and hammering the pig iron was slow and costly and took approximately 12 h to produce one ton of wrought iron. Henry Cort revolutionised this process by his invention of the puddling furnace, patented in 1783; this made it possible to produce fifteen tons of wrought iron in 12 h.

The puddling process resulted in a semi-solid 50 kg ball of mixed iron and slag, which was hammered and rolled into a bar, typically 25 mm square. This bar was classified as 'puddled bar' and was used as a starting point to manufacture higher quality wrought iron. 'Merchant bar' was produced by cutting puddled bar material into suitable lengths and piling the cut bars into orthogonal layers to produce a 0.5 m cube of piled iron bars bound together with wire or thin iron bar. This piled cube of iron was then reheated to 1,300°C and hammered and then rolled into bar. This process of piling, reheating, rehammering and rerolling was repeated to produce 'best bar', often used for engineering applications. Repeating the process again produced 'best best bar' typically used for chains, anchors and rivets. However, the 'best iron' from one district was equal to perhaps the 'best best iron' from another, so the above terms had a rather uncertain meaning except for the brands from reputable manufacturers, such as Best Staffordshire.

The strength and ductility of the wrought iron is improved each time the piling, reheating, rehammering and rerolling process is repeated. The slag stringers are broken up to produce a finer dispersion with the slag fibres principally aligned in the direction of final rolling. In good quality Victorian wrought iron, all evidence of the orthogonal orientation of the slag stringers contained in the differently orientated bars in the pile is totally eliminated during reheating, rehammering and final rolling of the bar to the next quality. However, some banding of the matrix structure may still be evident. The tensile strength of wrought iron can also vary significantly depending on where and when it was made, however reasonable confidence can be placed on the results of limited tensile strength testing if the elongation at fracture is more than 10%. Wrought iron is almost pure iron, malleable and ductile but less resistant than cast iron to corrosion.

Cort's further invention of grooved rollers, patented in 1784, paved the way for rolled structural sections such as flats, angles and tees (which were rare before 1820). Prior to this date, structural sections were generally hand wrought. Another significant development was the introduction of steam power to drive forge hammers in 1784 which enabled a larger mass of wrought iron to be formed to larger components. Wrought iron bulb channels and very small I-sections became available circa 1840s, initially used for deck beams and frames of iron ships. The world's first ocean-going propeller-driven iron ship was the SS Great Britain designed by Isambard Brunel and launched in 1843.

The challenge faced by Robert Stephenson of bridging the River Conway and Menai Straits for the Chester and Holyhead Railway in 1845, led to an outstanding advance in the use of wrought iron and probably one of the greatest technical achievements in engineering design. The spans contemplated were 122 m and 140 m, respectively, and hitherto the longest wrought iron span designed by Stephenson had been only 15.2 m. For the Britannia Bridge, any form of arch was not favoured by shipping interests, so Stephenson favoured a very stiff suspension bridge with the trains running through deep boxes of wrought iron and supported by chains on each side. With the help of William Fairburn and Eaton Hodgkinson, a chainless and continuous rectangular wrought iron box design evolved from extensive theoretical studies and experimental work. The design investigated buckling, bending, deflection, friction of riveted connections, fatigue, wind forces and temperature effects and thus wrought iron was established on a more rational basis than any other material. The Conway Bridge was completed in 1848 and load tests were used to prove the design and the Britannia Bridge was completed 2 years later, see Fig. 9.2.

Riveted plate girders and rolled I-sections gained popularity in the period

Fig. 9.2 Britannia Bridge, Menai Straits. Floating operations for the second box, circa 1850

from 1845 to 1855 and, by circa 1880, a large variety of rolled I-beams ranging in depth from 75 mm to 350 mm were available from manufacturers both in Britain and elsewhere. At this time, wrought iron beams encased in concrete were used for short span bridges. The use of wrought iron for major bridges declined rapidly in the period from 1876 to 1884. The Ashtabula Howe truss bridge (USA) collapsed in 1876 due to the derailment of a train in a snow storm and 2 years later the Tay lattice-girder bridge collapsed in a force 11 storm. The Tay bridge collapsed due to poor quality castings, unsuitable pinned and wedged joints and lack of understanding of wind loading on the high girders, leading to inadequate design strength. The last major bridge to be constructed of wrought iron was the Garabit lattice-truss and arch viaduct (France), designed by Gustav Eiffel and completed in 1884. Very little wrought iron was used after 1890, although there is evidence of some use of wrought iron sections as late as 1910.

Wrought iron continues to be manufactured in small but commercial quantities at a historic plant re-erected in Ironbridge Museum, Shropshire.

Steel

Structural steel sections were available in very limited sizes from 1850 onwards. The process of steel making was speeded up considerably by Henry Bessemer's converter (patented in 1856 and 1860) and Charles and Frederick Siemens' open hearth process (patented in 1861). However, the quantity of steel produced from the early furnaces was limited by the fact that the furnaces were lined with silica bricks (known as the 'acid' process) and only low phosphoric iron could be used successfully. Most of the native ores in Britain were phosphoric and therefore low phosphoric iron ore had to be imported from Spain and Sweden. In 1878, Sidney Thomas and Percy Gilcrest invented the 'basic' process, using bricks containing magnesia or dolomite and, by adding lime to the molten mix, they were able to remove phosphorus. The rapid exploitation of this process benefited USA and Germany more than Britain, each of which had huge resources of phosphoric iron ores and, by 1880, the world price for steel had dropped by 75% and steel suddenly became competitive with wrought iron. By 1887, Dorman Long and Company had produced a range of ninety-nine beam sizes, as well as a vast range of channel and angle shapes. By 1890 nearly all beams and other structural shapes were manufactured in steel. Developments in the manufacture and in the control of quality and strength of steel are reflected in British Standards from 1903.

The first major bridges to use steel as the principal structural material were the St. Louis and Brooklyn bridges in the USA. Steel was used for the main arch ribs on the St. Louis Bridge which was completed in 1874. The Brooklyn Bridge was under construction at the same time and steel was used for the cable stays and bridge deck trusses. The Firth of Forth Rail Bridge in Scotland was completed in 1890 and it represented the first large-scale use of open hearth steel in bridge construction (58,000 tons), see Fig. 9.3. It was then the largest spanning bridge in the world, being 30 m longer than the Brooklyn Bridge and it established the cantilever truss bridge as a serious alternative to the suspension bridge (see Chapter 10).

Steel derives its mechanical properties from a combination of chemical

Fig. 9.3 Forth Railway Bridge, Scotland. The first major steel bridge in Britain, 1890

composition, heat treatment and manufacturing processes. Steel is basically iron with carbon (typically 2% maximum by weight) and other elements added which can have a marked effect upon the type and properties of steel. These elements also produce a different response when the material is subjected to heat treatments involving cooling at a prescribed rate from a particular peak temperature. The manufacturing process may involve combinations of heat treatment and mechanical working which are of critical importance in understanding the subsequent performance of steels and what can be done satisfactorily with the material after the basic manufacturing process. Traditionally the molten steel was tapped from the furnace into ladles and then poured into large moulds to produce ingots. These ingots would normally be allowed to solidify and cool before reprocessing at a later stage by rolling. In modern steel making, the common types of furnaces are the basic oxygen furnace and the electric arc furnace. Molten steel is usually poured at a steady rate into a mould to form a continuous solid strand from which lengths of semi-finished product are cut for subsequent processing, thus eliminating reheating and first stage rolling of ingots. It should be noted that structural steel produced before 1955 was of much poorer quality than present day mild steel. Therefore it should be closely inspected for laminations, inclusions and deformities. Also, steel produced by the Bessemer or open hearth processes which were used up to the 1970s tended to be more brittle because it contained nitrogen from the air blast. The versatility of steel for structural applications rests on the fact that it can be readily supplied at a relatively cheap price in a wide range of different product forms, and with a useful range of material properties. Structural steel has a carbon content of approximately 0.10–0.25% and is therefore ductile with good fracture toughness and weldability. Although steel can be made to a wide range of strengths, it generally behaves as an elastic material with a high and relatively constant value of elastic modulus up to the yield or proof strength.

The age of steel opened the door to tremendous advances in long-span bridge building technology. The following bridges are notable examples in the period 1900 to 1950: Hell Gate arch bridge (USA, 1916), Quebec cantilever truss bridge (Canada, 1917), Sydney Harbour arch bridge (Australia, 1932), Elbe plate girder bridge (Germany, 1936), Golden Gate suspension bridge (USA, 1937).

The historical background to metal bridges and the significance of the developments in manufacturing processes for cast iron, wrought iron and steel are summarised in Fig. 9.4.

Some examples of iron and steel bridges, with brief comments on their maintenance histories are given in Table 9.3 at the end of this chapter.

COMPONENTS

Cast iron beams

Cast iron beams were purpose made for a particular application, usually with asymmetric flanges to take account of the low tensile strength, see Fig. 9.5. Other typical features were integral stiffeners, hog-backed top flanges and tapered flanges when viewed in plan. Bolt holes were formed in the castings where required and castings became more complex, limited by physical size. Cast iron beams were commonly used in composite bridge decks in conjunction with brick jack arches. Alternatively cast iron deck plates were used to span between the top flanges of the beams. These plates usually had upstand or downstand ribs or a curved profile.

Cast iron columns

Cast iron columns were usually of hollow circular construction and they were generally used in conjunction with a wrought iron crossbeam to form bridge piers. Columns with a 'X' cross section were more common in buildings than bridges. Cast iron columns usually had an integral head to support the beams above and the base of the column was sometimes cast separately. Cast iron was also used for piled bridge foundations, especially on swing bridges (see Chapter 11).

Wrought iron beams

Wrought iron beams generally have a symmetrical cross-section and were thinner than cast iron ones in relation to overall dimensions. The maximum depth of rolled I-sections was generally 200 mm, with exceptions up to 350 mm, and therefore larger beam sections were fabricated from plates, angles and tees with riveted connections. In some instances girder webs were fabricated from strips of wrought iron plate to form a lattice. Holes for rivets or bolts were generally drilled but sometimes punched.

Steel beams

Prior to the issue of British Standard 4 (BS4) in 1903, the shapes and sizes of beams were settled by individual manufacturers. BS4 listed thirty British

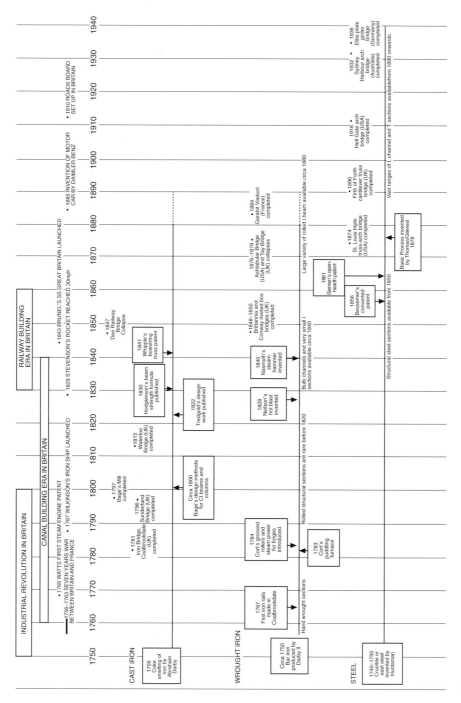

Fig. 9.4 Historical background to metal bridges

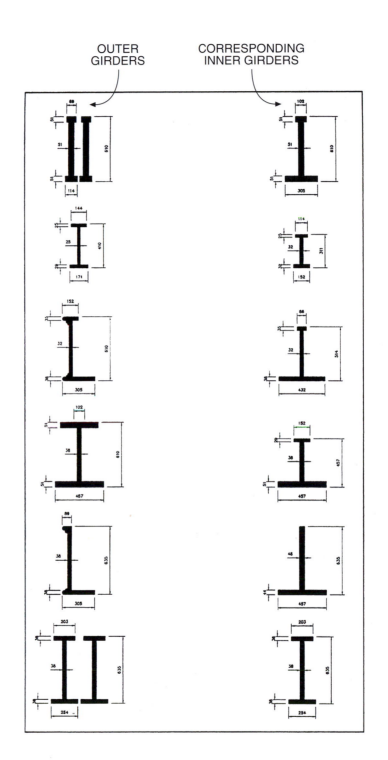

Fig. 9.5   Typical cast iron beam cross-sections

Standard Beam sizes (Ref Nos BSB 1 to 30). BS4 was reissued in 1921 with new beam reference numbers NBSB 1 to 18 and British Standard Heavy Beams and Pillars (Ref Nos NBSHB 1 to 11). BS4 was reissued in 1932 listing forty British Standard Beam Sizes (Ref Nos BSB 101 to 140). Continental beams were also exported to Britain in the period 1900–1930. Universal beams were introduced in 1959 (see Bates, 1990).

Steel columns

Standard stanchion sizes were first introduced by BS4 in the 1932 version and the stanchions comprised single or double beams with additional flange plates. Universal columns were introduced in 1959 (Bates, 1990).

Hybrid beams

It was not uncommon in the nineteenth century for beams to be constructed of a hybrid of either cast iron and wrought iron or steel and wrought iron. Possible examples are as follows:

- wrought iron bars cast into the bottom flange of a cast iron beam
- cast iron top flange/wrought iron bottom flange
- cast iron web/wrought iron flanges
- cast iron decorative features/wrought iron structural elements
- cast iron beams/wrought iron ties under bottom flange
- wrought iron beams with additional steel flange plates

Once painted, these combinations may look monolithic, however the implications can be significant as discussed under 'Management Practice'.

Hybrid beam decks

It is not uncommon to find that bridge decks have been widened or strengthened by the addition of newer or reused beams and therefore a hybrid population of cast iron, wrought iron and steel beams may exist in the same deck. It is important to appreciate that cast iron beams reach peak load at much smaller deflections than wrought iron/steel beams and for a given strain (or deflection) cast iron beams can be stressed to their limit, while more ductile beams could be stressed to only a fraction of their potential. This feature needs to be considered carefully in a load assessment of such a deck.

Arches

The earliest metal arch bridges were made from cast iron arched ribs, cast in sections and connected to form a complete member. The connections reflected carpentry practice and were either mortise and tenon, dovetail or bolted joints. It is a common occurrence for structures exploiting new materials for the first time to use the construction techniques of the previous technology. For shorter spans, the arched ribs had solid webs and the section height decreased towards midspan. For longer spans open spandrels, typically with circular features, were used between the bottom arched rib and deck level. To make the arched ribs stronger, open lattice construction was used either in the form of a lattice girder or lattice box. Also, a combination of cast iron and wrought iron was used for compression and tension members, respectively. Both materials were used to good effect in tied arch bowstring bridges. Typical examples of metal arches are shown in Fig. 9.6.

One of the earliest examples of a steel arch is the curiously named Iron Bridge, Rothbury, built between 1870 and 1875. When the Tyne Bridge, Newcastle, was constructed in 1928, this two-pin steel arch was the longest of its type in Britain, with a span of 162 m. This record is currently held by the Runcorn–Widnes Road Bridge, which was opened in 1961, with a span of 330 m.

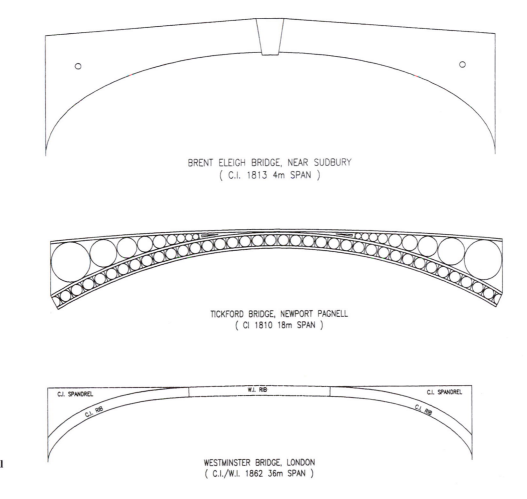

BRENT ELEIGH BRIDGE, NEAR SUDBURY
(C.I. 1813 4m SPAN)

TICKFORD BRIDGE, NEWPORT PAGNELL
(CI 1810 18m SPAN)

WESTMINSTER BRIDGE, LONDON
(C.I./W.I. 1862 36m SPAN)

Fig. 9.6 Typical examples of early metal arches

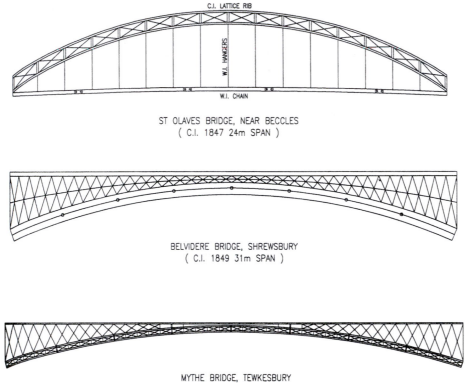

ST OLAVES BRIDGE, NEAR BECCLES
(C.I. 1847 24m SPAN)

BELVIDERE BRIDGE, SHREWSBURY
(C.I. 1849 31m SPAN)

MYTHE BRIDGE, TEWKESBURY
(C.I. 1825 52m SPAN)

Fig. 9.6   Continued

Jack arch decks

Brick jack arches springing from cast iron/wrought iron or steel beams was a common form of construction for railway overbridges in the nineteenth century. A study of jack arch bridge decks has revealed that there are principal families of these bridges with specific configurations, for which it appears that there were empirical design rules for cast iron bridges and 'design standards' for wrought iron/steel bridges.

The vast majority of bridge decks in current service with cast iron beams and brick jack arches have the following design features:

- two ring jack arches with a circular arc profile, springing from the lower flanges of the beams
- structural backing (e.g. concrete, gypsum lime or mortared masonry) to at least the extrados of the arch and more likely to the top of the beams
- a minimum tie area of approximately 260 mm^2/m length of beam in the outer bay of transverse spanning jack arches, at springing level.

The vast majority of bridge decks with wrought iron/steel beams and brick jack arches have the following design features:

- two ring jack arches with a circular arc profile, generally springing from

the lower flanges of the beams and sometimes springing from part way up the beams, typically supported on angles riveted to the webs
- some structural backing
- a minimum tie area as for cast iron beam construction
- a maximum aspect ratio for a jack arch of 10 (beam spacing/rise of the arch).

In some cases, mass concrete has been used for the jack arches, with or without permanent metal plate formwork.

Longitudinal spanning jack arches are often found in conjunction with through deck construction (see below).

Through decks

This form of construction is common for bridges under and over railways where shallow construction depth is required below the running surface. The longitudinal edge girders are typically made from wrought iron or steel with cross girders or trough floors spanning between the edge girders, forming a half-through deck. For longer spans the edge girders are usually replaced by trusses or box girders. For spans in excess of 50 m, overhead cross-bracing is usually provided between the top of the edge members, forming a through deck.

Box girder decks

Early box girder decks are rare and the Conway Railway Bridge is a fine example of a pair of riveted wrought iron tubes through which a single track passes in each direction. Steel box girder decks were pioneered during and after World War II and many were constructed in Germany in the 1950s and 1960s, with spans up to 280 m.

PERFORMANCE

Cast iron

One of the greatest virtues of cast iron is its high resistance to corrosion and testament to this is the number of cast iron bridges still giving good service today. The Iron Bridge was closed to vehicular traffic in 1931 due to limitations of strength. However, the Coalport Bridge built nearby in 1799, still carries traffic. The style and decoration of many cast iron bridges also makes them desirable to keep.

As cast iron is brittle, it is essential to consider carefully the implications of any defect in the cast iron that has arisen either during the casting and construction or as a result of subsequent deterioration and modification. These defects can manifest themselves as notches, pitting, holes, voids and cracks. Surface pitting and holes below the surface at the top of the mould are caused in inadequate ventilation of mould gases. Voids or open texture at the web/flange and web/stiffener junctions are caused by the slower cooling of the middle of the section compared to the surface of the mould.

Fig. 9.7 Uckfield cast iron bridge collapsed in 1903. The vehicle causing the collapse was owned by the local authority

Such defects may act as local stress raisers and could cause cracking in the tension zone.

Cast iron is also vulnerable to cracking caused by impact and fatigue loading, see Fig. 9.7. This was recognised in the Iron Commissioners' report dated 1849 which recommended that a factor of safety of 6 should be applied to the live load for railway bridge design after the collapse of the Dee Railway Bridge in 1847 (Fig. 9.8). Longitudinal fractures to beam flanges can also occur if work to services buried in the bridge deck is not carried out by appropriate methods. Impact resistance is poor and reduces gradually with falling temperature and with increasing phosphorus content. However, cast iron has a low sensitivity to fatigue if permissible stresses are kept within reasonable limits and it also has considerable damping qualities.

For cast iron beams used in masonry jack arch decks, it is particularly important that adequate horizontal restraint is provided to the edge beam to prevent lateral bending and/or twist in the beam. Jack arch construction is discussed in more detail below.

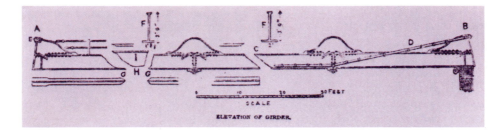

Fig. 9.8 Elevation of failed trussed girder from the Dee Railway Bridge (1847), showing position of cracks

Wrought iron

Wrought iron is less resistant than cast iron to corrosion but has greater ductility and resistance to fatigue compared to steel. Deterioration usually results from water penetration through the bridge deck and in the steam blast zone on the soffits of railway overbridges. The following areas are particularly vulnerable to corrosion:

- bottom flanges and rivets
- bottom of web stiffeners
- webs of large plate girder edge beams at level of verge/footway
- other areas where moisture and dirt have become entrapped (e.g. between plates)
- splash zones.

Severe corrosion of wrought iron is characterised by a laminated or stringy longitudinal texture, see Fig. 9.9. This is due to its fibrous structure which is a result of its manufacture and serves as a useful indicator to distinguish between wrought iron and steel.

The traditional method of joining structural wrought iron members is the rivet. It is important to appreciate that rivets produce a non-slip connection, whilst ordinary bolts in clearance holes produce a connection with a certain amount of slip.

Steel

All metal bridges require an effective paint system to prevent general corrosion. However, this has a greater significance for steel bridges compared

Fig. 9.9 Excessive corrosion to a wrought iron beam. Note laminated structure of piece of corroded metal resting on bottom flange

to cast iron and wrought iron. The vast majority of problems with steel bridges result from a breakdown of the paint system and the onset of corrosion due to lack of regular maintenance and the vulnerable areas are similar to those listed for wrought iron. Steel corrodes more readily than either cast or wrought iron. Exposed steel in a rural inland environment usually corrodes at a rate of less than 0.05 mm loss of parent metal per annum and, in an industrial environment, the corrosion rate is usually 0.05–0.10 mm per annum. It should be noted that heavy accumulations of rust do not necessarily indicate significant loss of section as general rust occupies up to ten times the volume of the original material.

Early steel members were jointed using rivets and by circa 1900, structural steelwork began to be welded by the oxyacetylene process. The first welded steel bridge was built in Britain in 1931. Welded fabrications dated later than 1930–1935 are almost certainly to be electric arc welded. Site fabrications were generally bolted to overcome the difficulties in quality control and inspection of welded connections.

Brittle fracture is feared by engineers as it can occur suddenly and without warning, for example Hasselt Bridge, a welded steel structure in Belgium, collapsed overnight without any traffic loading. It is alternatively named cleavage fracture as the actual fracture surface, when new, has a characteristic crystalline appearance like sugar. Brittle fractures are normal to the direction of stress and exhibit negligible deformation, and fracture is initiated at a defect such as a mechanical notch, casting defect, weld defect or corrosion pit, see Fig. 9.10. Steel exhibits a brittle–ductile transition temperature, where fractures at temperatures below the transition are brittle. Early steels tended to have high transitions but successive improvements in metallurgy have lowered the value. Factors that affect the transition include notch acuity of the defect and rate of loading. Transition temperatures were formerly measured by laboratory tests such as Charpy impact

Fig. 9.10 Example of brittle fracture initiated at a defect

and Tipper notch tension, named after their inventors. Nowadays the more fundamental fracture toughness tests are employed.

Fatigue cracking, like brittle fracture, is initiated at defects. Cracks propagate progressively, depending on the frequency of loading, so that overall rates can be rather low, for example cracks in the welds of steel bridges can take months to propagate a metre or so. This is helpful as it usually enables cracks to be detected before there is a serious problem. In the 1970s and 1980s, there was a spate of fatigue problems in welded bridges, see Fig. 9.11. In North America, cracks occurred in fabricated girders at the ends of cover plates welded to bottom flanges and at the connections between vertical web stiffeners and bottom flanges.

In Europe, there have been few problems of this type due to different design philosophies and a greater awareness of fatigue in the early days of designing for welded construction. On the other hand there has been a number of fatigue problems in welded orthotropic steel decks (so-called battle decks) constructed in the 1950s and 1960s in Germany, France, Holland and Britain. The cracks have occurred at welds between longitudinal stiffeners and deck plate, and between longitudinal stiffeners and crossbeams. Fatigue problems can also occur in other types of connection, for example from rivet holes in plated or truss construction and in hanger cables at entry points into anchorages.

Fatigue most commonly occurs under repeated tensile-loading, but for

Fig. 9.11 Cracking at a welded connection due to fatigue

welded connections it can also occur under pulsating-compression. The latter is not generally known and is due to the zone of residual tensile stress adjacent to a weld, which results in locally repeated tension.

Design and assessment of fatigue is carried out using S–N curves (also termed Wöhler curves) of stress range against number of cycles to failure, where the curves have been generated from relevant data and analysed as mean minus two standard deviations. Design S–N curves are provided for different types of connection in Standards such as BS5400, Steel Concrete and Composite Bridges.

Jack arch construction

Most of the structural deficiencies in jack arches occur in the outer bays of transverse spanning jack arches as a result of inadequate lateral restraint being provided. Where this is the case, the distress usually consists of either cracking or deformation of the arch barrel or more likely missing or loose brickwork at the crown of the arch and these are indicators of some lateral movement or rotation of the outer beams. It is very unusual to find structural defects to inner bays because each jack arch provides lateral support to the adjacent bays. However, defects may be present in inner bays as a result of movement of the substructure. The end bays of longitudinal spanning jack arches do not generally suffer from lack of restraint because some restraint to lateral movement is usually provided by the substructure.

Lateral restraint in the outer bays of transverse spanning jack arches is usually provided in the form of tie rods or bars. In the case of service bays, the edge beam may be adequately restrained by a steel plate or strap connected to the bottom flange of beams on each side. (However, a cast iron plate just sitting on the bottom flanges will not provide adequate restraint.) The ties are vulnerable to corrosion where they pass through the jack arch and they may be found damaged or missing.

For cast iron bridges with transverse spanning jack arches, it appears that the design of these decks was based on empirical rules where a minimum area of ties was provided. As nearly all of these decks have concrete backing or other competent structural backing, it is considered that these decks behave similarly to filler beam decks and the ties are necessary to ensure that this composite action can be utilised. For the vast majority of these decks, ties are located at the level of the bottom flange. However, in the case of wrought iron/steel structures with transverse spanning jack arches, composite action does not appear to have been utilised in their design.

The beams supporting the jack arches are susceptible to lateral bending and/or twist if there is inadequate horizontal restraint and this is of particular concern for cast iron beams. Beams are also susceptible to corrosion due to deck leakage and external factors such as steam blast.

Half-through girder decks

In half-through girder decks, lateral torsional stability of the main edge girders is provided by the crossbeams connecting the edge girders at the bottom flange level. Due to the shape of this arrangement of elements, it is commonly referred to as a U-frame. The effectiveness of U-frame action will depend on the transverse vertical bending stiffness of the edge girders and the rotational stiffness provided by the connection between the edge girder and the deck. This effectiveness can be impaired by corrosion to stiffeners and connections.

Increases in live loading over time can produce imperfections such as twist and out-of-straightness which should be measured and taken into account in any load assessment.

CONSERVATION

Methods of conservation are discussed below and summarised in Table 9.1.

Cracking and fractures

Cracks and fractures in cast iron can be repaired by either welding or stitching. Welding of cast iron needs expert knowledge and it is advisable to consult specialist organisations before considering such work. It should be

Table 9.1. Methods of conservation

Type of damage	Scheme	Comment
Cracking/fractures	Cold stitching of castings.	Specialist technique which can restore a large proportion of original strength.
	Welding:	
	Cast Iron.	Specialist technique requiring control of thermal stress and distortion. Rarely used in structural situations.
	Wrought Iron.	Specialist technique similar to CI. Laminar tearing can occur adjacent to new welds.
	Steel.	Techniques well known, expertise widely available.
Loss of section due to corrosion	Deal with source of problem and stop further corrosion.	Problems rare with CI due to high resistance to corrosion.
	Replace member like-for-like.	Maintains appearance. Use traditional materials and techniques where possible.
	Weld or rivet new plates to member.	Care required to minimise any change in appearance.
	Replace rivets.	Replacement rivets are more authentic than HSFG bolts.
Deformation	Heat treatment.	Specialist technique to restore original shape. Undergoing trials.

noted that the welding of cast iron requires special attention to the selection of electrodes and the preheating and post-heating of the parent metal and therefore it is not easily controllable under site conditions. Although welding of cast iron is rarely used in structural situations, strong reliable repairs were achieved in the spandrel rings on the iron bridge, Stratfield Saye, Hampshire, built in 1802. Here the largest member was just over 50 mm square and the bridge which is located in a landscaped park, was restored to its original 2 tonne capacity. Braising of cast iron is another option for non-structural members.

Welding of wrought iron is possible although, as with cast iron, it should be approached with caution and only after obtaining specialist advice. It is important to appreciate that the laminar structure of wrought iron makes it susceptible to laminar tearing adjacent to a new weld.

The technique of welding steel is well known and expertise is widely available. Steel can be welded to wrought iron, as shown in Fig. 9.12. However, in this repair, migration of inclusions occurred from the wrought iron to the steel.

An alternative technique for repairing cracked cast iron and wrought iron/steel castings is cold stitching. Proprietary systems such as the Metalock process can be virtually invisible, it can restore a large proportion of the original strength and control of thermal stresses and distortion is not applicable. The Metalock process consists of fitting special nickel alloy keys into drilled apertures across the fracture, see Fig. 9.13. The high strength and highly ductile keys are peened into a metal-to-metal condition and become almost integral with the parent metal. Between the keys, overlapping studs are installed along the fracture line. Final peening and hand-dressing of the surface completes the repair, see Fig. 9.14.

Fig. 9.12 A welded repair between wrought iron (top left) and steel (top right). Note inclusions on right hand side of v-shaped weld

Fig. 9.13 A stitch repair to a cast iron beam.
Drilling of apertures to receive alloy keys

Corrosion

In conjunction with the conservation of any corroded metal members and components, it is important to rectify the cause of the corrosion to prevent further deterioration. Significant loss of section to cast iron is rare due to its high resistance to corrosion. The most common approach to reinstating

Fig. 9.14 A stitch repair to a cast iron beam.
Completed repair highlighted in white chalk marks

the loss of section on wrought iron and steel members is to bolt new steel plates or sections to the affected member, which unfortunately often leads to an unsightly repair. A preferred approach is to cut out the defective area and weld or rivet a new section in place, see Fig. 9.15, thus maintaining the original thickness of the section.

Rivets may need to be replaced if they are severely corroded or if they have to be removed to secure a new plate, preferably utilising the existing rivet holes. In all cases, rivets should be removed by a pneumatic 'rivet buster', see Fig. 9.16, which shears off the head and punches out the shank. Oxyacetylene cutting torches should not be used due to the risk of damage to the original hole.

Old rivets should be replaced with steel rivets 'riveted up hot', which fill the old rivet hole, see Fig. 9.17. The use of HSFG bolts to replace rivets is not good conservation practice because the connection is less effective and not authentic.

If severe loss of section has occurred to a member then it may be aesthetically and/or economically advantageous to replace the affected member. Where possible, traditional materials and techniques should be employed. For example, specialist organisations can produce authentic wrought iron

Fig. 9.15 Welded repair to wrought iron beam. Magnetic particle testing of completed weld

Fig. 9.16 Removal of rivets using a pneumatic 'rivet buster'

members by the puddling process to close tolerances and exacting technical specifications and damaged sections or complete cast iron members can be recast using the original as a pattern, see Fig. 9.18. It should be noted that cast iron shrinks when cooled and therefore the mould needs to be oversized. In some instances it may be possible to remove a corroded edge girder, transfer an identical inner girder to the edge of the deck and install a new inner girder, thus overcoming the problem of reproducing an exact replica and maintaining the appearance of the bridge.

Deformation

A new heat treatment technique developed in the USA has been evaluated recently by the Highways Agency in trials to straighten a damaged steel bridge beam spanning the M5 motorway in Somerset. The bridge was struck by a tipper truck which caused the outer 840 mm × 290 mm universal beam to displace by 400 mm over a 2.5 m length. To repair the beam, oxyacetylene torches were used to heat the lower flange to between 500°C and 600°C in a V-shaped pattern. Horizontal jacks were used to restrain expansion away from the original line so that as the beam cooled, internal stresses developed which gradually straightened the beam. The trials were successful and the technique has been used on a second bridge as a cost-effective alternative to replacing the damaged section of beam. Research in the USA is said to have shown that, providing local steel temperatures do not exceed 750°C, the strength of the straightened beam is virtually identical to its original value.

Fig. 9.17 Hot set riveting on a wrought iron member. Rivets are set red hot using a pneumatic gun and 'holder-up'

STRENGTHENING

Various methods of strengthening old bridge superstructures have been developed as described in the following sections and summarised in Table 9.2.

Main beams

One solution to under-strength main beams is to insert additional main members in between existing members and the new members may be designed to carry part or all of the live loading. Ideally, the additional members, should be visually similar to the original members but this is not always possible. In certain circumstances, it may be acceptable to construct additional beams in reinforced concrete or prestressed concrete, with a soffit profile similar to the existing members. This approach has been used for the strengthening of cast iron arched rib decks, especially over rivers, examples of which are the Trent Bridge, Nottingham, and Town Bridge, Thetford. Alternatively, universal beams have been inserted between the cast iron arched ribs on St. Saviour's Bridge, Ottery St Mary, which is a sched-

Fig. 9.18 Damaged
section of cast iron
spandrel arch recast and
stitched into position

uled ancient monument, and arched steel beams have been inserted
between the cast iron ribs on Brandy Wharf Bridge, Lincolnshire. In the
latter case, the original cast iron arch carries no live load.

A similar approach was used for the strengthening of the grade II listed
Farringdon Street Bridge in London, constructed in 1867, where a structural
steelwork grillage with a concrete deck was designed to fit in between the
cast iron main beams. The new steelwork and deck was supported on
reinforced concrete extensions to the existing granite columns and was
designed to act independently of the cast iron beams, which effectively
became non-load-bearing decorative features, see Fig. 9.19. The pier extensions were cast to the same profile as the original pier top castings and a
bitumen bonded cork joint was formed between the concrete and the cast
iron beams. The original cast iron bracing was then refitted to the pier extensions. This strengthening solution obscured little of the cast iron members
and also permitted phased traffic management on the road above.

The original main girders/beams can be replaced with similar steel members of a higher capacity and this method was used for the strengthening
of the grade II listed Adelaide Bridge in Royal Leamington Spa, constructed
in 1891. Two of the original steel inner arched lattice girders were moved

Table 9.2. Methods of strengthening

Structural member	Strengthening scheme	Comment
Understrength main girders/beams	Insert additional main members or steelwork grillage between existing members.	New members may carry all live loads.
	Replace internal members and retain edge members.	Live load transfer to edge members can be eliminated
	Weld or bolt additional plates to member or encase in concrete.	Care required to minimise any change in appearance.
	Plate bonding using steel or CFRP plates.	Relatively simple, does not detract from visual appearance and minimises disruption to traffic.
	Ensure composite action between main members and concrete infill.	Entails replacing existing infill and possibly new shear connectors.
	Replace existing decking and/or infill with lighter material.	Can cause significant disruption to traffic.
	Reduce span.	Difficult to minimise visual impact. Propping not desirable.
Understrength secondary members	Insert additional secondary members.	New bolted or bonded connections may be required.
	Replace members.	Similar appearance can be maintained.
	Weld or bolt additional plates to members.	Can be achieved without significantly affecting appearance.
	Plate bonding using steel or CFRP plates.	See comment for main girders/beams.
	Make members non-load bearing.	Has been achieved by constructing new deck slab.
	Add splints to existing members.	Void between splint and existing member can be filled with grout.
	Add shear connectors and infill to trough decks.	Composite action can be achieved with minimal change in appearance.
	Add/replace ties in external bays of jack arches.	Particularly important for cast iron main beams.
Understrength deck construction	Replace deck construction with new RC slab.	Economical, but can be obtrusive.
	Replace non-structural fill with concrete.	Possible to achieve a composite deck and retain jack arches/deck plates.

Fig. 9.19 Farringdon Street Bridge, London. A new structural steelwork grillage with a concrete deck installed between the old cast iron main beams

to the edge of the deck to maintain the visual appearance of the bridge and five new inner girders were installed. A pinned connection between the edge girder and adjacent inner girder was made to prevent load transfer to the edge girder and loading on the footway above the edge girder was carried by new transverse box sections cantilevered from the top of the inner girders. For the strengthening of Queen's Avenue Bridge over the Basingstoke Canal, built in 1899, the seven wrought iron plate girders were in poor condition and were replaced by twelve rolled steel beams. In order to create the original plated appearance of the main beams, 6,000 imitation rivet heads were welded to the new beams. In this case, it is arguable whether or not the use of false rivets is good practice from a conservation viewpoint. However, it is commendable that a significant effort has been made to make the new beams visually similar to the original beams.

The assessment of Lambeth Bridge, London, in 1995 revealed that the inner steel arched girders were deficient at midspan due to high horizontal shear stresses at the top and bottom of the web. The solution adopted was to bolt 20 mm thick steel cover plates to the web plates and flange angles, utilising the existing rivet holes, see Fig. 9.20. A precise installation sequence was devised which allowed rivets to be removed while keeping the bridge

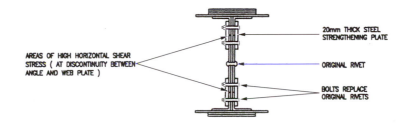

Fig. 9.20 Lambeth Bridge. Section through strengthened inner arch rib

AREAS OF HIGH HORIZONTAL SHEAR STRESS (AT DISCONTINUITY BETWEEN ANGLE AND WEB PLATE)

20mm THICK STEEL STRENGTHENING PLATE

ORIGINAL RIVET

BOLTS REPLACE ORIGINAL RIVETS

open to traffic. Similarly, bolted or welded steel plates can be fixed to wrought iron or steel beam webs and flanges to enhance shear capacity and bending capacity, respectively.

In the case of Underbridge No. 194, constructed circa 1890s to carry the Warrington to Manchester railway line over the Manchester Ship Canal, a massive plating exercise was carried out to strengthen the main span lattice girders without changing the visual appearance of the girders. Unfortunately, a new deck and overhead bracing were required on the live half of the main span in conjunction with new approach spans to replace the original plate girder decks, to accommodate modern loading and increased train speed. However, as the clearance over the canal is 23 m, the change in construction is not obvious to the casual observer.

For other through girder decks, additional stiffeners can be fixed to edge beams and the connection with transverse members can be improved by increasing the rigidity of the connection, using additional plate and/or shear connectors fixed to the edge beam and cast into a reinforced concrete slab. With all such work, careful consideration needs to be given to maintaining the visual appearance of the bridge.

The solution adopted to strengthen the final compression member local to the pinned base on Warburton Bridge over the Manchester Ship Canal was to encase the member in concrete, see Fig. 9.21. The cantilever design was unique at the time of construction, circa 1890s, and the same design was used 25 years later for the Quebec Bridge over the St. Lawrence River which collapsed during construction due to a buckling failure of the final compression member. Careful detailing is required with concrete encasement to prevent water traps between the concrete and metalwork. A high standard of workmanship has been achieved on Warburton Bridge and the scale of the repair is small in relation to the overall structure, both factors contributing to a minimal change in appearance of the structure. More extensive concrete encasement of the three inner arched ribs was undertaken on Waterloo Bridge, Betwys-y-Coed, in 1923. This historic cast iron bridge, constructed by Thomas Telford in 1815, carries the A5 London to Holyhead road and, although the strengthening work is visible from below, the concrete has performed well and is not visible on the elevations, see Fig. 9.22.

In contrast, the strengthening of the cast iron road bridge in Fig. 9.23 has been carried out very unsympathetically. Apart from the obtrusive propping, the edge beam on the left-hand span appears to have been replaced by a reinforced concrete beam which is now spalling along the soffit arris, see Fig. 9.24.

The drastic solution adopted for the road bridge in Fig. 9.25 was to replace all the cast iron main beams with pretensioned concrete beams, retaining only the edge beams. Examples of strengthening such as the two described above are far too common and are usually driven by minimum cost. Hopefully a greater appreciation of our engineering heritage and awareness of alternative strengthening methods, such as plate bonding, will lead to more innovative strengthening of existing main members in the future, rather than propping or replacement.

Plate bonding techniques, more commonly used to strengthen reinforced/prestressed concrete members, have been developed to

Fig. 9.21 Warburton Bridge. Concrete encasement to prevent buckling failure

strengthen cast iron girders. Bures Bridge in Suffolk was strengthened in 1991 using steel plates bonded to the soffit of the five cast iron arched girders. Prior to the site work, design assumptions and a satisfactory working technique were verified by a series of full scale trials and load capacity tests performed on a cast iron beam recovered from a similar bridge. Clamps on the edges of the bottom flange were utilised to apply pressure during adhesive curing. Although the clamps were designed only as a temporary measure, they were left in place as a permanent fixture as a precaution against future debonding. The steel plates were bolted to the bottom flange using HSFG bolts at each end of the bridge. This technique is relatively simple, does not detract from the visual appearance of the bridge and minimises disruption to traffic.

A further development of plate bonding techniques is the use of carbon fibre-reinforced plastic (CFRP) plates. CFRP plates have the advantage of being lighter and stronger than steel plates, which makes them easier to install and the required thickness of plate is reduced. The required thickness of plate can be further reduced by prestressing as used on Hythe Bridge in 1998/1999. Prior to site work, feasibility trials were undertaken on three cast iron beams of similar cross-section saved from a demolished

Fig. 9.22 Waterloo Bridge, Betwys-y-Coed. Concrete encasement on inner arched ribs

Fig. 9.23 Unsympathetic strengthening by propping and installing a concrete replacement edge beam

structure. A prestressing device was developed such that the end anchorages could be clamped to the bottom flanges of the beams eliminating the need to drill into the cast iron. The trials were successful and provided the necessary data for the strengthening works. The degree of prestress was designed to remove all tensile stresses from the cast iron under 40 tonne loading. This was achieved using four CFRP plates per beam, each stressed

Fig. 9.24   Spalling to concrete replacement edge beam. Note tapered bottom flanges to cast iron main beams

Fig. 9.25   Inner cast iron beams replaced with pre-tensioned concrete beams

to a total of 18 tonne. Allowance was made for beam restraint and local effects, examined by finite element analysis. The prestressing of cast iron members in this way does raise the issue of potential fatigue cracking which needs careful consideration and it would be wise to carefully consider the loss of the prestress due to possible relaxation of the resin during hot weather.

A further application of composites has been used in the strengthening of Tickford Bridge, Newport Pagnell, see Fig. 9.26. This cast iron bridge is a scheduled ancient monument and the three largest spandrel rings and the lower main chord members have been strengthened using a resin impregnated carbon fibre sheet system, see Figs 9.27 and 9.28. The system was applied in layers, up to a maximum of fourteen layers, with a maximum total thickness of 10 mm. The design used a fracture mechanics approach to analyse the effects of any stress concentrations at joints or cracks and the fracture toughness of the adhesive layer was determined from laboratory testing.

It may be possible to reduce the level of overstress in the main girders/beams by reducing the weight of the deck construction. Lightweight concrete has been used to replace existing infill and decking material. Lightweight glass reinforced plastic deck panels supported on new steel cross girders have been used in the strengthening of High Bridge, Staffordshire, which has now been reopened to pedestrian traffic.

Generally, for older composite bridge decks comprising metal main girders and a concrete top slab, no positive shear connection was provided. Therefore, it is usual to assume that no composite action takes place for the initial assessments. However, a significant hidden strength may exist if

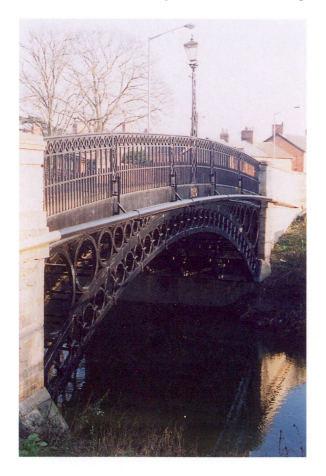

Fig. 9.26 Tickford Bridge, Newport Pagnell. Constructed in cast iron in 1810

Fig. 9.27 Application of resin impregnated carbon fibre sheet system to spandrel rings

composite action could be guaranteed. In the case of Westminster Bridge in London, see Fig. 9.29, the results of the load testing and post-test computer modelling demonstrated that the capacity of the main cast iron/wrought iron arched ribs had a potential increase from 7.5 tonnes to 40 tonnes due to composite action. Such action was found to increase the bending capacity of the wrought iron ribs at midspan and tensile load was shed from the relatively weak top sections of the cast iron rib spandrels into the concrete deck near the piers. To guarantee composite action in service, the strengthening work involved installing special shear connectors bolted to the original cast iron/wrought iron ribs and replacing the original concrete and timber infill with a reinforced lightweight concrete deck.

Sometimes it is not feasible to strengthen a bridge by carrying out *in situ* strengthening. This was the case for grade II listed Bedminster Old Bridge in Bristol, built in 1883. The main members comprised double web wrought iron lattice girders with abutting flanges and access for any work was severely restricted due to the close web spacing and numerous services. The solution adopted was reconstruction of the bridge deck, retaining only the

Fig. 9.28 Resin impregnated carbon fibre sheet system bonded to lower main chord. Note local build-up of resin around transverse member

Fig. 9.29 Westminster Bridge, London. Main inner girders comprise wrought iron ribs over the central section and cast iron spandrel ribs at the ends

edge girders and parapets. The new deck comprised steel I-beams fitted between the services, acting compositely with a concrete deck slab. To maintain a continuous soffit, flat GRP enclosure panels were installed under the new beams. The edge girders and parapets were refurbished and supported off the new hidden edge girders.

The reconstruction of Stephenson's Britannia Bridge following the fire in

1970, is an extreme case where it was considered to be a practical impossibility to rebuild the bridge with an outward appearance closely similar to the original design. The fire caused cracks up to 290 mm wide in the top of the tubes at the towers, tilting and displacement of the cast iron bearings and maximum midspan deflections of 700 mm, see Fig. 9.30. The two main spans were in imminent danger of collapse, so Bailey bridge units were erected in the original jacking slots in the three towers to provide temporary vertical support. The tubes were beyond repair and the dilemma then faced by the designers for the reconstruction work was the desire to replace the tubes with new tubes jacked up the piers as adopted by Stephenson, the need to demolish the old tubes in a safe manner and the problem of restoring rail traffic on the London–Holyhead line as soon as possible. To remove the tubes in full span lengths would have required the use of at least two of the world's largest floating cranes, operating in waters where tidal currents were likely to cause additional problems and expense. Alternatively, to reverse the original method of erection, i.e. to lower the tubes by jacking, was considered to be too great a risk, considering their damaged condition. The conclusion reached was the design for the new spans had to be capable of wholly supporting the Stephenson tubes so that these could be cut up and removed in comparatively small sections. The possibility of maintaining the original appearance of the bridge by constructing replacement tubes was investigated and rejected because the lifting slots in the towers were fully occupied by the emergency supports. The final design selected was

Fig. 9.30 Britannia Bridge, Menai Strait. Fire damage to the original wrought iron tubes in 1970

triple spandrel-braced steel arches for the main spans and continuous steel box girders supported on circular concrete columns for the approach spans. The masonry towers and abutments were not seriously affected by the fire and were therefore retained. The new design permitted one of the Stephenson tubes to be supported by the new steelwork and used for rail traffic, whilst the other was dismantled and replaced by a new rail deck. Also, provision was made for the inclusion of an upper road deck carrying a diversion of the A5 London–Holyhead trunk road in order to relieve the loading and congestion on Telford's Menai Bridge. Fortunately Stephenson's other great tubular bridge at Conway is still in service, although the span was reduced by 27 m in 1899, see Fig. 9.31.

Secondary members

There are many options available to overcome the problem of under-strength secondary members compared to main girders/beams, because the secondary members are generally less visible. Several options were used for the strengthening of the cast iron grade II listed Mythe Bridge, near Tewkesbury, constructed by Telford in 1826. The shear strength of the transverse cross-beams was enhanced by bonding 20 mm thick steel plates to the web of the cross-beams and the longitudinal cross-bracing at each end of the bridge was strengthened using 8 mm thick angle splints and the voids between the splints and cross-bracing was filled with flowable cementitious grout, see Fig. 9.32. Additional transverse bracing was installed along the centre spine of the arch, with one end bolted to steel sleeves, in turn epoxy bonded to the original transverse tie beams and the other end bolted to the top of the arch cross-beams. The bridge now has a 17 tonne capacity to

Fig. 9.31 Conway Bridge, Conway. The pair of wrought iron tubes were strengthened in 1899 by constructing new piers and reducing the span by 27 m

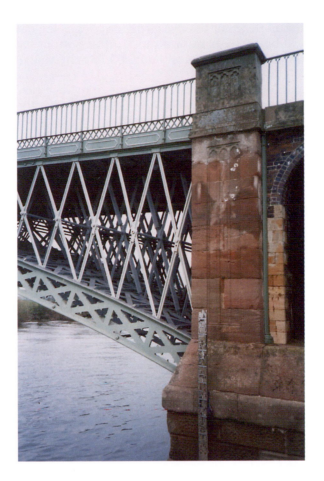

Fig. 9.32 Mythe Bridge, near Tewkesbury. The splints on the longitudinal cross-bracing are just visible above the first inner arched rib

allow the Tewkesbury fire brigade to provide cover for villages on the other side of the river.

The heritage value of secondary members should not be overlooked, just because they are less visible. On Westminster Bridge mentioned earlier, the weak buckle plates and cross-beams were effectively made redundant for load carrying by the new reinforced concrete slab, but they were left in place to preserve as much of the original structure as possible. Other options include replacing the original member with a steel member of higher capacity and replacing the existing decking and/or infill with a lighter material. For trough decks, additional members can be fitted in the troughs, encased in a new reinforced concrete slab. In the case of Aireworth Bridge, Bradford, the secondary members comprised steel Hobson sections (which have a similar profile to brick jack arches), supported on an inverted T section. These sections spanned transversely between plate girder edge beams. It was possible to strengthen this bridge to 40 tonne loading utilising composite action by fixing HSFB bolts to the Hobson sections and casting a reinforced concrete slab over the top. For jack arch construction, if the effective area of the tie bars/rods in the external bays of transverse spanning jack arches is less than 260 mm^2/m length of beam, then the capacity of the deck is likely to be below optimum. It is usually straightforward to provide adequate horizontal restraint to edge girders by replacing corroded tie bars and/or fixing additional tie bars or straps to the bottom flange. It is often possible to carry out these works without spoiling the visual appearance of the structure.

In certain situations, it may be advantageous to replace cast iron members with replica members made from SG iron, thus achieving the same appearance, higher strength, greater toughness and better shock resistance. These improved properties are achieved by the addition of approximately 0.04% of magnesium or cerium to molten cast iron, which causes the graphite to form into small nodules when the metal solidifies. SG iron was used for the replacement stanchions on the ends of the cross-beams on Brunel's Clifton Suspension Bridge, Bristol, see Fig. 9.33, and for the replacement parapets on Adelaide Bridge and Westminster Bridge mentioned earlier.

Decks

The original metal deck plates spanning between the top flanges of girders/beams with non-structural pavement material above, rarely meet current highway loading standards. On Tickford Bridge mentioned earlier,

Fig. 9.33 Clifton Suspension Bridge, Bristol. Spheroidal graphite cast iron used for the replacement stanchions on the end of the cross-beams

the problem of fractured cast iron deck plates was solved in 1900 by the addition of wrought iron arched plates, bolted through the original plates close to the ribs and in 1976, a 300 mm thick reinforced concrete slab was laid over the deck on a 20 mm thick cushion of closed cell polyethylene foam to avoid hard local contact with the cast iron beams. The concrete slab also permitted better distribution of wheel loads, thus reducing the maximum load on any one arched rib. For the refurbishment of three cast iron canal footbridges at Farmers Bridge Junction, Birmingham, the deck plates were blast cleaned and refitted, and steel angles were bonded to the outer edge of the decks and the inner face of the edge beams to form a structural connection. A reinforced concrete slab was then cast on top of the deck plates providing further lateral stability.

A replacement concrete deck can be an economical strengthening solution and it also provides an ideal surface to support a waterproofing membrane, thus minimising the risk of future corrosion to the metalwork. However, care is required to minimise the visual impact, as illustrated in Fig. 9.34 Here, a cast iron arched rib bridge has been strengthened by adding a reinforced concrete deck and infill between the arched ribs. It could be argued that the appearance of the bridge has now dramatically changed, however, it has been possible to retain the original main members and the arch profile has been maintained by replicating the same profile on the underside of the concrete string course.

Girder decks with either brick jack arches or deck plates spanning from the bottom flanges of the girders can be strengthened by replacing the fill with concrete to achieve a composite deck. The jack arches or deck plates then become nonstructural and can often be preserved. Unfortunately, this was not the case for the bridge shown in Fig. 9.35. Here, the concrete fill

Fig. 9.34 A reinforced concrete deck and infill between cast iron arched ribs

Fig. 9.35 **Strengthening of the cast iron girder deck with infill concrete.** Unfortunately, the original jack arches or deck plates have been replaced with permanent steel formwork

has been cast on top of permanent steel formwork. Service pipes on both elevations spoil the appearance of this bridge.

PIERS AND SUBSTRUCTURES

Metal piers on multispan bridges are not very common and are mainly found on railway bridges. The performances of metal piers are generally related to cracking of cast iron and corrosion of wrought iron and steel. Important considerations in any load assessment and subsequent strengthening are the effects of any eccentric vertical loading and the effect of traction and skidding loads transmitted from the superstructure. Refurbishment and strengthening techniques will be similar to those discussed above. The cracked cast iron columns on Hungerford Rail Bridge, London, built in 1862, have been repaired over the full height of 10 m by cold stitching, see Fig. 9.36.

Movement of substructures can cause problems in the superstructures of metal bridges. Cracking along the crown of transversely spanning jack arches may be due to differential settlement. Cracking of the iron work on the famous Iron Bridge was found to be caused by the abutments moving towards each other and deterioration of the north abutment. (The south abutment was replaced in 1800 by one much smaller, farther up the valley side and connected to the main arch by two approach spans.) Remedial works on the north abutment consisted of excavating the filling material and constructing a hollow reinforced concrete box inside the masonry structure. The new walls provided stability to the external masonry and precast beams supported a concealed concrete deck and acted as struts to resist earth pressures. The second phase of the restoration was the construction

Fig. 9.36 Cast iron piers of Hungerford Rail Bridge. Crack repairs using cold stitching

of a U-shaped reinforced concrete strut below the river bed between the abutments.

MANAGEMENT PRACTICE

Inspection

In order to carry out a successful rehabilitation or strengthening scheme, it is necessary to undertake a thorough inspection of the bridge. Older metal bridges present peculiar features which are not found in modern bridges, therefore it is important that the inspector is aware of these as outlined below.

The paint systems on metal bridges can hide the fact that members can be made up of a combination of different metals and possibly timber. Bolts, screws and rivets were often cleverly hidden. Likely combinations are detailed in the section on hybrid beams.

When it is required to identify the material it is helpful to be aware that cast iron has a granular surface, whereas wrought iron and steel have smoother surfaces and the only certain way to distinguish between wrought iron and steel is by metallographic examination. However, wrought iron tends to have a more stringy appearance than steel often visible in areas of significant corrosion or at the point of fracture. The fracture surface of cast iron is granular and the cracks tend to be straight or slightly jagged. The date of bridge construction can be misleading as metal components were frequently reused.

A close inspection should reveal whether a structural member could have been rolled or not. If it has a constant cross-section throughout its length, it is most likely of wrought iron or steel, but if it has integral stiffeners or variable web height/flange width which would preclude rolling, it is certainly cast iron. Sometimes it is possible to identify mould lines formed on the surface of cast iron members. Members with hammered surfaces or forged ends tend to be wrought iron. Members with curved surfaces, such as columns and fluting, could not be forged and are therefore likely to be made from cast iron.

Riveted plates and connections were used for both wrought iron and steel. Rivets are prone to corrosion and should be examined for signs of movement under load. False rivets made of mould putty have been found and can be easily distinguished by the tap of a hammer. Virtually all welded ferrous metal will be steel, except in the case of obvious repairs by welding.

When measuring the thickness of cast iron columns, misleading results can occur due to misplaced cores. Therefore, it is good practice to drill three small holes 120° apart, but not on the same horizontal plane, and take the mean of the three measurements.

When inspecting cast iron members, particular attention should be given to the identification of notches, pitting, holes and voids which could act as a local stress raisers. Any lateral bending or twisting should give cause for concern (see Performance). Wrought iron members should be carefully examined for laminations, inclusions and deformities.

When inspecting half-through girder decks, it is important that any imperfections such as twist and out-of-straightness are measured. By using measured imperfections, a more realistic assessment of live load capacity can be achieved.

Cracks can often be identified visually by rust staining from the freshly exposed surfaces or by simply seeing the crack itself if it is wide enough to be discernible. Various types of commercially available inspection equipment can be used which is based on magnetic or ultrasonic effects. Impact-echo (hammer tapping) is a traditional method of detecting cracked members which has worked well over the years.

Maintenance

Painting is the single most important maintenance activity (as opposed to repair work) in iron and steel bridges. Moreover, it has a strong interaction with the requirements of good conservation, see also Chapter 13.

Paint and coating systems are used on metal bridges to provide protection from corrosion. Good surface preparation is critical to achieve the optimum performance from any paint/coating system, but before the method of surface preparation is selected, it is necessary to establish the extent of work involved in a site survey. The survey should identify the nature of the existing protection system, the extent of work and the health and safety hazards

Table 9.3. Some examples of iron and steel bridges

Bridge	Engineer	Date	Span(s) m	Comments
Iron Bridge	Abraham Darby	1781	30.5	The oldest surviving cast iron bridge. Side arches inserted in the stone abutments, 1921. Closed to vehicles, 1931. Abutments stabilised by reinforced concrete invert, 1973.
Stratfield Saye	Wilson	1802	12	Cast iron. Refurbished, cracks repaired by welding, restored to 2 tonne capacity, 1998.
Tickford, Newport Pagnell	Wilson and Pervis	1810	18	Fractured deck plate strengthened with buckle plates, 1900. A reinforced concrete slab laid on to the deck, 1976. The three largest spandrel rings and lower main chord members strengthened with CFRP in layers of up to 10 mm thick, 1999.
Waterloo, Betws-y-Coed	Telford	1815	46	Cast iron. Three inner arch ribs encased in concrete, 1923.
Cleveland, Bath	Hazzledine	1833	33.5	Cast iron. Strengthened by insertion of reinforced concrete beams between edge beams, 1930.
Britannia	Stephenson	1850	70–140–140–70	Twin rectangular riveted box girders of wrought iron. Seriously damaged by fire in 1970. Rehabilitated in a different structural form with little regard for conservation.

Table 9.3. Continued

Bridge	Engineer	Date	Span(s) m	Comments
Forth Rail	Baker and Fowler	1890	520–520	Cantilever and suspended span trusses of steel. In continual use. A gigantic structure.
Tyne, Newcastle	Mott, Hay and Anderson	1928	162	2-pin steel arch. The longest of its type at the time of construction.
Lambeth, London	Humphreys	1932	38–45–50–45–38	Steel arch structure. Strengthened by bolting 20 mm steel cover plates to web plates and flange angles using original rivet holes.

to workers, public and environment. Measurement of paint thickness can be carried out using commercially available equipment and the measurements can be used to identify vulnerable areas which may fail in the short term.

Hazardous ingredients in old paint systems are mainly red lead and zinc chromate pigments, commonly found in primers. It is therefore important when taking paint samples for analysis to remove paint flakes down to the substrate. If analysis identifies hazardous components in the coating, the selection of surface preparation techniques assumes a greater importance. Where total removal of the existing paint becomes necessary, consideration will need to be given to containment systems, surface preparation filtration systems, environmental sampling, debris sampling, respirators for workers and worker health monitoring and appropriate disposal of hazardous waste by certified contractors. On bridges where successive layers of paint have been applied over the years, there can be a significant increase in dead weight as discussed in Chapter 11 on movable bridges.

There is a wide range of paint systems available to suit differing environments, access arrangements and maintenance intervals. These systems are continually evolving with advances in paint technology, changes in environmental legislation and results of performance. In recent years, high build coating systems have been developed to achieve at least the same or better durability in fewer coats than their five or six coat predecessors. Thermally sprayed metallic coatings of either zinc or aluminium, combined with a paint system have many advantages over conventional full paint systems and have therefore become popular. Alternative coating systems include elastomeric urethane, glass flake coatings and water-borne coatings.

Paint and coating systems also enhance the appearance of the bridge by the appropriate choice of colour for the finish coat. For older bridges, it is desirable to use colours that closely match the colours that would have been available at the time of construction. If possible, the original colour scheme should be reinstated, based on archive information or paint sampling. This is historically correct and good conservation practice, but the colours are often drab and uninteresting. For bridges located in conservation areas or bridges with listed status, the colour will need to be agreed with local planning authorities. Different colours can be used to good effect to highlight structural action, for example stiffeners painted in a darker colour/shade than the parent plate. Also, in environments where heavy contamination of surfaces is likely, light colours should be avoided as these rapidly become discoloured. Conversely, light colours may aid visual detection of fractures or cracks, particularly in cast iron. Historic structures traditionally had a matt/silk finish coat, however this finish tends to hold more dirt compared to a gloss finish. The colour and type of finish therefore need careful consideration on a bridge-specific basis.

BIBLIOGRAPHY

Angus, H.T. (1976) *Cast Iron: Physical and Engineering Properties*. 2nd edn. London: Butterworths.

Ashurst, J. *et al.* (1991) *Practical Building Conservation. English Heritage Technical Handbook, Vol. 4 Metals*. Gower Technical Press.

Bates, W. (1990) *Historical Structural Steelwork Handbook*. The British Constructional Steelwork Association Ltd, London.

Blackwall, A. (1985) *Historic Bridges in Shropshire*. Shropshire Libraries.

Chettoe, C.S. *et al.* (1944) 'The Strength of Cast Iron Girder Bridges'. *J. Instn. Civ. Engrs*, **22**(8).

Commissioners Appointed to Inquire into the Application of Iron to Railway Structures. (1849). Report, (2 Volumes). Leeds: HMSO.

Davies, H.E. *et al.* (1982) *The Testing of Engineering Materials*. 4th edn. McGraw-Hill.

Doran, D.K. (ed.). (1992) *Construction Materials Reference Book*. Butterworth Heinemann.

Matheson, E. (1873) *Works in Iron-Bridge and Roof Structures*. Spon.

Morgan, J. (1999) 'The Strength of Victorian Wrought Iron'. *Proc. Instn. Civ. Engrs, Struct. and Bldgs*, vol 134.

Sutherland, R.J.M. (1985) *Recognition and Appraisal of Ferrous Metal*. Proceeding of Conference on Building Appraisal, Maintenance and Preservation. Bath, July.

Sutherland, R.J.M. (ed.). (1997) *Structural Iron 1750–1850*. Studies in the History of Civil Engineering, Volume 9. Ashgate Valiorum.

Swailes, T. (1996) '19th Century Cast Iron Beams: Their Design, Manufacture and Reliability'. *Proc. Instn. Civ. Engrs, Civ. Eng*. Vol. 114.

Chapter 10

Suspended bridges

GLOSSARY

Suspended bridge	Generic term for suspension, cable-stayed or hybrid bridge.
Suspension bridge	Bridge having its superstructure suspended from main cables.
Cable-stayed bridge	Bridge having its superstructure supported by stay-cables attached to towers.
Fan-stays	Stay-cables radiating from tops of the towers in a fan shape.
Harp-stays	Stay-cables parallel to one another in a harp shape.
Hybrid bridge	Bridge having both main cables and stay-cables.
Main cable, catenary cable	Often used as a generic term for the principle suspensive element of a suspension bridge. Chain or wire-cable.
Stay-cable	Supportive element of a cable-stay bridge, usually wire cable.
Chain	Suspensive element used in early bridges c1800 to 1880. Composed of wrought iron links and pins.
Wire cable, rope	Suspensive element composed of wires, early versions used wrought iron, later versions used cold drawn steel.
Hanger, suspender	Suspensive element connecting deck to main cables, early versions were wrought iron rods, later versions were cables.

Tower, pylon, pillar	Supportive element in cable-stayed and suspension bridges.
Span length	Length of deck between towers or support points. Ultra long-span refers to lengths exceeding 500 m, long-span refers to 200 m to 500 m.
Blacksmith bridges	Term used by Ruddock to describe bridges of lightweight iron construction built in Scotland and Ireland between 1816 and 1834. Relevant to bridges of similar construction built in this era elsewhere.
Simplistic footbridges	Lightweight suspension footbridges constructed by journeymen making best use of available materials and maintained likewise without concern for retaining originality.

BACKGROUND

In this chapter, methods of conservation of the different types of suspended bridge are addressed. Particular emphasis is given to suspended footbridges because they are more numerous and tend to receive less attention than the longer span highway bridges. Methods of conservation are considered in relation to the main components, i.e. towers, catenary chains and cables, hangers and stays, and decks.

Suspended bridges are interesting structures that attract public attention and affection. Many are sited in scenic locations, where they stand alone and can be seen at their best. It is therefore important that they are properly looked after, particularly the lesser known bridges and footbridges that can sometimes be overlooked. There are few suspended bridges that do not have a degree of historic significance and therefore merit conservation.

Types of suspended bridge

The main type of suspended bridge can be categorised according to the system of suspension, i.e. catenary (the classical suspension bridge), cable-stayed, and various combinations of the two, as shown in Table 10.1. In Britain, engineers became seriously interested in suspended bridges in the early 1800s. Some of the earliest structures were cable-stayed having a fan configuration. For the longer spans constructed in the 1820s and later, designers of the time such as Telford, Brown, Tierney Clark and others invariably adopted the catenary system of suspension. When maintenance or strengthening is required for these better known bridges it has to be carried out in the full glare of publicity and care has to be taken to ensure that originality is maintained. On the other hand the lesser known suspended bridges and footbridges often have to be maintained on a lower budget and maintenance engineers would benefit from knowledge of their

Table 10.1.   Common configurations of suspended bridges

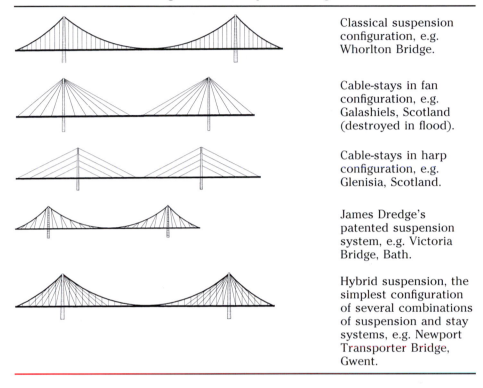

	Classical suspension configuration, e.g. Whorlton Bridge.
	Cable-stays in fan configuration, e.g. Galashiels, Scotland (destroyed in flood).
	Cable-stays in harp configuration, e.g. Glenisia, Scotland.
	James Dredge's patented suspension system, e.g. Victoria Bridge, Bath.
	Hybrid suspension, the simplest configuration of several combinations of suspension and stay systems, e.g. Newport Transporter Bridge, Gwent.

design and construction. This is particularly relevant to footbridges as there is a large number of structures having historic significance but treated at a low level of importance.

Performance

Early suspended bridges could be regarded as being rather fragile in comparison with, say, masonry arches. There are no reserves of hidden strength that can be identified by clever methods of analysis and if overloaded, suspended bridges can, and do, collapse without warning, examples are given in later sections of this chapter. They are prone to vibration and the attendant problems, also addressed later.

The early suspended bridges were constructed in cast iron and wrought iron which are materials much less susceptible to corrosion than modern high strength steel. In consequence, most of the original ironwork has survived in remarkably good condition despite receiving, in some cases, only minimal maintenance. There are of course the exceptions where components are located in a particularly aggressive environment.

The early suspended bridges invariably had timber decks. Nor surprisingly, these have had to be replaced at regular intervals placing a burden on maintenance. Over the years many have been changed to materials more durable than timber, examples are given in the section on decks.

Overall, suspended bridges have performed surprisingly well, as evidenced by the large numbers that have survived, some reconstructed, some strengthened or otherwise modified, and some in original condition. Some examples of surviving suspended bridges, up to 180 years old, are listed in Tables 10.8 to 10.10.

HIGHWAY BRIDGES

Catenary suspension

Early suspension bridges were stricken with problems due to wind-induced oscillations and there were numerous cases of serious damage, for example Brighton Chain Pier was damaged in 1833, 1836 and in 1896 when it was totally destroyed, and the Menai Bridge was damaged during construction, twice in 1826, in 1836 and again in 1839. Wind remains a potential problem for surviving structures, for example Whorlton Bridge, 1831, a high level suspension bridge across the River Tees was damaged by wind in 1976.

These early designs had little stiffness and relied on the dead weight of the deck and the resulting tension in the main cables to resist the wind effects. It was quickly realised that a horizontal wind could cause small vertical oscillations that resonated and developed into large and destructive deflections. Engineers of the day tackled wind effects in different ways, one school of thought being to 'stiffen' the suspension system. James Dredge built several bridges, starting with Victoria Bridge across the River Avon at Bath in 1836, using a patented design of inclined hangers. The bridge was refurbished in the 1940s and remains in its original form and open to pedestrians some 165 years after construction, see Fig. 10.1. This is a good performance for the time, albeit Dredge was fortunate in building on a sheltered site which would not experience the full effects of wind, and traffic was not unduly heavy.

More importantly, Rendel advocated the use of truss girders to stiffen the

Fig. 10.1 Victoria Bridge, Bath. The first of Dredge's bridges having his patented system of suspension. Inset shows one of the concreted anchorages

deck and stabilise behaviour under wind action. In repairing the Montrose Bridge in 1838, he added a 3 m deep wooden truss to the 132 m span to give a depth-to-span ratio of 1 to 43. Surprisingly, stiffening of the deck was not generally adopted in suspension bridge design until some years later.

Suspension bridges were also found to be prone to pedestrian-induced vibration. Broughton Bridge, a cast iron suspension structure of 44 m span constructed in 1828, collapsed in 1831 when sixty soldiers marched across causing resonant vibrations. The debris was subsequently inspected and it was reported that the principal fracture had taken place at a bolt in the main chain. The fracture was described as being granular without any fibrous appearance indicating that it was brittle. To this day, the Albert Bridge across the Thames in London carries a notice instructing troops to break step, see Fig. 10.2. Moreover, there are historic bridges still in use today that are very lively under pedestrian loading.

Albert Bridge, built in 1873, now has a hybrid system of suspension but was originally designed by Ordish as a cantilever bridge having each half supported by sixteen stays radiating from the tops of the towers as fans. Dredge reported that catenary cables, each composed of 1,000 wires of 2.5 mm diameter had been fitted from the tops of the towers with ties connected to the stays (rods) to prevent them from sagging. After a time, cracks developed in the protective covering to the cables and the wires showed signs of corrosion. This led to the cables being replaced by chains in 1887, after only 14 years' service. At around this time Bazalgette carried out strengthening work described at the time as being to 'make the bridge more like a conventional suspension bridge'.

In the 1840s, Stephenson evaluated different structural forms when

Fig. 10.2 Notice displayed on Albert Bridge, London. This became very topical in June 2000 when the Millennium Bridge downstream was opened and almost immediately closed due to excessive oscillation under pedestrian loading

designing the rail crossing of the Menai Straits. He concluded that an adequately stiffened suspension bridge would be too expensive and proceeded with the design of the stiffened box girder Britannia Bridge, one of the most significant bridges of the nineteenth century. In making this decision, Stephenson seems to have discouraged further development of long-span suspension bridges in Britain and few highway suspension bridges were constructed in the next 75 years, as evidenced in Tables 10.8 and 10.9. Clifton Bridge is an exception, albeit it was designed in 1831, but not completed until 1864. Medium-span bridges, such as the 129 m span Hammersmith Bridge (1887), continued to be built.

In 1937, Chelsea Bridge was constructed across the River Thames in London with stiffened steel plate girders, articulated twin steel towers and a composite deck. This remains in its original form after some 60 years. Two ultra long-span bridges were built in the 1960s. The Forth Road Bridge (1964) has a deep truss, orthotropic steel deck and 38 mm hand-laid mastic asphalt running surface. The Severn Bridge (1966) has an aerodynamic steel box girder, inclined hanger cables to provide added damping and an orthotropic steel deck surfaced with 38 mm hand-laid mastic asphalt.

After the Humber Bridge was built in 1980, no more suspension highway bridges have been built in Britain as it became feasible to have cable-stayed bridges which are easier to construct and more economical than suspension bridges of the same length.

Some examples of surviving British suspension bridges are given at the end of this chapter in Table 10.8.

Cable-stayed

One of the early cable-stayed bridges was at Craithie (1834) built for vehicular use, but downgraded to pedestrian use in 1857 and subsequently renewed in 1885. The Albion Bridge across the River Avon at Twerton was built by Motley in 1837. This had stay-cables (rods) in the harp configuration fixed to twin towers at each end. The rods were inter-connected by vertical hangers to provide stability in a similar manner to the scheme adopted retrospectively on the second Severn crossing. Unfortunately there is little information about the bridge's performance and it was dismantled in 1879. Haughs of Drimmie Bridge in Scotland (1837) has remained in service having received minimal maintenance and continues to be open to light traffic.

Other engineers designed hybrid suspension systems having main cables and a combination of inclined hangers and stays, for example Arnodin's Transporter Bridge in Newport, South Wales, see Fig. 10.3.

In Germany, some elegant steel cable-stayed bridges were constructed in the 1950s and 1960s many of them across the River Rhine. These had lightweight orthotropic steel decks designed to be economic with material.

In Britain, interest in purely cable-stayed bridges was renewed 127 years after Twerton Bridge, when Charles Brown designed George Street Bridge in Newport having a main span of 152 m and completed in 1964, see Fig. 10.4. This was to be followed by many others including Wye Bridge (1966) which forms part of the first Severn crossing and has a main span of 235 m, and Erskine Bridge (1971) across the Upper Clyde which has a span of 305 m.

Fig. 10.3 Transporter Bridge, Newport, Gwent. Example of a hybrid suspension system

Fig. 10.4 George Street Bridge, Newport, Gwent

It is notable that the latter span length is similar to Tamar Suspension Bridge, built 12 years earlier.

Recent cable-stayed bridges have encountered dynamically lively behaviour as exemplified by the oscillations exhibited by some stay-cables at modest wind speeds and in the presence of rain. Whereas the earlier engineers attempted to deal with this by methods such as increasing stiffness of the superstructure or tying the superstructure down, engineers now set out to design aerodynamically stable cross-sections and use added damping to cables and decks when it turns out to be necessary.

Some examples of British cable-stayed and hybrid bridges are given at the end of this chapter in Table 10.9.

FOOTBRIDGES

Although there were few significant suspended road bridges constructed in Britain in the latter half of the nineteenth century, suspended footbridges, mostly in the classical form with catenary cables and vertical hangers, continued to be built and there are many remaining examples. Most are in the span range 30 m to 50 m and have performed well over the years, many having exceeded the current notional design life of 120 years. Unfortunately pedestrian bridges have tended to attract less attention than highway and rail bridges, and less is known about their construction and maintenance histories. Many have been quietly dismantled and replaced.

Early footbridges

Many of the early footbridges were cable-stayed, some having wire cables, for example Galashiels (1816) and Kings Meadow (1817). Ruddock has described these and some of the other bridges built in Scotland and Ireland between 1816 and 1834, as 'blacksmith bridges'; they were constructed in wrought iron and required blacksmith skills to fabricate (Ruddock, 1999). Some of them also featured the early use of wire cables. Footbridges of this era that have survived and are still usable include Kirkton Bridge of Glenisia having harp stays of 16 mm diameter wrought iron rods and iron portals at each end. A feature of these bridges is their extreme lightweight and elegant appearance. Decorative ironwork at the bases of the towers of Kirkton Bridge added balance to their appearance and may also have contributed some lateral stability. Two of the surviving 'blacksmith' bridges, Haughs of Drimmie (1837) and Craithie (1834) were built for vehicular use.

Gattonside Bridge (1826) across the River Tweed between Melrose and Gattonside, is a suspension bridge having a conventional configuration, with chains and vertical hangers of 13 mm diameter. The towers are of masonry and the main span is 91 m. Sadly, when the bridge was rehabilitated in 1991–1992 much of the elegant ironwork that had survived for 165 years was replaced.

Stowell Park footbridge, a privately owned structure across the Kennet and Avon Canal, was designed by James Dredge to his patent system of suspension and constructed c1840. It has survived in original condition having apparently required minimal maintenance beyond replacement of the timber deck from time to time, see Fig. 10.15. It is not open to the public.

Simplistic footbridges

There is a type of suspension footbridge which can best be described as simplistic. Such structures are invariably privately owned or ownership is unclaimed. In consequence, performances and states of repair are usually less that would be required of publicly owned structures and use by the general public is not encouraged. Simplistic footbridges are not to be confused with Ruddock's 'blacksmith bridges' constructed by craftsmen in the early 1800s to the contemporary state-of-art. In contrast, simplistic footbridges were constructed by journeymen making best use of available materials and have been maintained along the same lines.

Simplistic footbridges are invariably located in fairly remote parts of the countryside and have the following general characteristics:

- There is no longitudinal stiffening girder and superstructures are very flexible.
- Decks are composed of longitudinal timber planks typically having gaps between them and fixed to cross-pieces bolted to the hanger rods.
- Widths are narrow to the extent that there is insufficient room for people to cross simultaneously from opposite ends.
- Ad hoc maintenance over the years has reduced the amount of original material.
- There is no evidence that the structures were technically designed and dates of construction have rarely been recorded so that origins are unknown.
- Due to the great flexibility of the superstructures, simplistic bridges are very lively and unstable for timid or infirm people.

Burn Footbridge across the River Exe in Devon has a single span of 30m, see Fig. 10.5. It is a typical example of a simplistic suspension footbridge, believed to have been constructed in the mid to late 1800s associated with a railway line no longer in use. It has a wooden portal tower at one end and

Fig. 10.5 Burn Footbridge, Devon, c1900. Example of a simplistic footbridge having very lively dynamic behaviour

a more recently constructed steel tower at the other, see Fig. 10.6. The main cables are of circular section link chains and are also shown in Fig. 10.6. At one end, the chains are joined together several metres beyond the wooden tower and a stranded wire cable of 19 mm diameter completes the connection to a substantial concrete anchorage.

Protective wrapping around the tie-back cable has been rubbed and damaged by grazing cattle. At the other end, the chains were previously tied to a tree, but are now anchored conventionally. The hangers are rods of 16 mm diameter looped over the chains and bolted to cross-rods at deck level. Longitudinal wooden planks are bolted to the cross-beams to form the 560 mm wide deck. A handrail of 13 mm diameter is positioned on either side.

The bridge is very lively being prone to simultaneous horizontal (sway) and vertical oscillations. It also has a disconcerting tendency to 'crab', posing an added threat to pedestrians. At the time the bridge was inspected, there was no significant corrosion and it could be considered to be in relatively good condition, if the unconventional aspects can be ignored.

Backs Wood Footbridge, also across the River Exe, is a simplistic structure having two spans of 10 m and 16 m, see Fig. 10.7. It has three portal towers 2.4 m high and constructed of bolted steel beams. The towers are of more recent construction than the rest of the bridge and do not match the size

Fig. 10.6 Burn Footbridge. Steel tower erected in recent times to replace a tree which had originally been used. The cradle and suspension chain are shown in the inset

Fig. 10.7 Backs Wood Footbridge. The main cables are supported by pulley wheels on the towers and hangar rods are freely looped over the cables

of the central masonry river piers, see Fig. 10.8. The three masonry piers have been capped with concrete such that two of the towers are bolted in place and the third has its base partly encased in concrete.

The main cables are wire ropes of 50 mm diameter passing over pulley wheels acting as cradles and fixed to the tops of the towers, as shown in Fig. 10.7. At the time of inspection, there were several broken wires but these were isolated occurrences and the cables were otherwise in surprisingly good condition.

Hanger rods of 13 mm diameter are freely looped over the main cable where they are held in place longitudinally by friction alone. This connection although seemingly insecure has the advantage that detritus is unlikely to settle and encourage corrosion, as occurs in fixed connections. It is an example of a case where it is best to let well alone. At their bottom ends the hangers are bolted to cross-rods. The deck is composed of longitudinal timber planks fixed on to the cross-rods, and is 560 mm wide so that, like Burn Footbridge, there is room for only one person to cross at a time. A handrail of 13 mm diameter is positioned on either side of the deck. Several of the hangers had been bent out of shape, presumed to be the work of vandals.

As with Burn Footbridge, the structure is very lively, having first bending and lateral sway frequencies of around 1 Hz. Walking frequencies for the average person are in the range of 1.5 to 2.5 Hz, but the natural reaction when crossing is to slow down due to the rickety appearance of the bridge, and 'lock-on' to the bridge frequency. This worsens the oscillations. In contrast, when crossed at a normal pace rate the bridge is much less lively, but this requires a degree of courage. The structure could be stiffened if the deck were fixed to the central pier, the original fixings having become ineffective. At the time it was inspected, there was little corrosion.

Fig. 10.8   Backs Wood Footbridge. The central pier is out of scale with the superstructure and may have been built for an earlier and larger bridge

A simplistic footbridge in the USA, shown in Fig. 10.9, has concrete portal towers and main cables of stranded wire rope, as defined in Table 10.6. Hanger bars are looped over the cables and locked more positively than Backs Wood Bridge. The timber plank deck is arranged in a similar manner, but is in need of repair as shown in Fig. 10.9.

Although unsuitable to be used by the general public, and having lost much of their originality, simplistic footbridges are typically at least 100 years old and have heritage value and a certain charm that is worth retaining. However, even in their as-built condition, simplistic footbridges would have been hazardous in relation to the standards of safety required nowadays.

Later footbridges

Teddington Lock Footbridge, constructed in 1889 but modified some years later, is typical of the next stage in the development of suspension footbridges, see Fig. 10.10. The towers are of riveted iron frames having their legs encased in concrete and their cross-beams exposed. The main cables are composed of twin wire ropes kept apart by the hanger brackets, the hangers are steel rods. The nearside end span shown in Fig. 10.10 has an

Fig. 10.9 Simplistic footbridge in the USA. Surprisingly similar to the Exe footbridges

unusual feature with the footway skewed to the direction of the main span. There are additional stay-cables, also skewed. In 2001, the original main cables and hangars were still in place, having survived 112 years' service. Substantial lattice parapets contribute stiffness to the superstructure, nevertheless it can be excited and is regarded locally as being lively.

Gaol Ferry Footbridge, Bristol, constructed by David Rowell in 1935 is of lattice construction, see Fig. 10.11. There are similarities to the Teddington Lock Footbridge in that the main cables are twin wire ropes on each side having similar hanger brackets and hanger rods. It has given excellent service over 65 years' service life and required no significant refurbishment beyond the normal cycles of repainting.

In contrast to the good performance of Gaol Ferry Footbridge, Trews Weir Footbridge, Exeter, constructed in 1935 and of similar design, has given less satisfactory service. Some of the hanger rods fractured due to a combination of corrosion and high stresses, and were replaced by stainless steel in 1984. Most of the trusses and steelwork were renewed in 1993, leaving the towers, catenary cables and anchorages in their original form. The footbridge has a reputation for lively behaviour and 'children love to get it going'. This plus its nearness to the sea and the consequential presence of chlorides may account for the deterioration.

Burgate Footbridge in Hampshire is one of the few suspension footbridges

Fig. 10.10   Teddington Lock Footbridge, 1889

Fig. 10.11   Gaol Ferry Footbridge, Bristol 1935. Insets show attachment of hanger cable to catenary cable and one of the main anchorages

to be built in Britain since the 1930s. It was designed by E.W.H. Gifford and used recycled elements from Bailey bridging. The structure is economic and elegant, representing an advanced level of simplicity and economy that has not been equalled in recent times. It remains in its original condition having required no significant refurbishment over 52 years of admittedly light service in a rural environment, see Fig. 10.12.

Some examples of suspended footbridges are given at the end of this chapter in Table 10.10.

Fig. 10.12 Burgate Footbridge, Hampshire. An economic and elegant design that has not been equalled in recent times. Inset shows connections between catenary cable, hanger cable and parapet rail

CONSERVATION OF TOWERS

The towers for catenary suspension bridges were sometimes single pillars, but more commonly portal frames at each end of the suspended span. There were different systems of articulation, as categorised by Pugsley (1968):

(1) Stiff towers fixed at their bases and having saddles at their tops that permit the main cables to slide in a span-wise direction
(2) Towers hinged at their bases to permit rotation in the span-wise direction and having the main cables fixed at their tops, ie acting as simple struts.
(3) Towers fixed at their bases and having the main cables fixed at their tops, i.e. acting as vertical cantilevers.

The older bridges usually had type (1) towers and were constructed in masonry, cast iron or wrought iron. Type (2) towers, hinged at their bases, are a more recent development made possible by plated steel construction used earlier by French engineers. Long-span bridges of the twentieth century have usually been type (3).

The principal forms of loading are vertical compression, wind and longitudinal movement of the main cables. On occasions when saddle bearings at the tops of the towers become seized, movement of the superstructure has to be accommodated by rocking of the towers. In practice this has rarely caused problems as the towers, particularly masonry ones, are sufficiently flexible to be able to cope with the movements.

Masonry

Masonry towers have performed well as they have usually been constructed of good quality stone and there are many surviving examples in excellent condition having required minimal maintenance over the years. There have been cases where rendering has been applied, but records have been unclear as to whether this was an original architectural finish, or a retrospective addition. Maintenance is best carried out using the methods developed for other heritage masonry structures, e.g. maintaining drainage facilities, replacing defective stone and repointing lime mortar joints when appropriate.

On one of the masonry portal towers of Wheeling Bridge, West Virginia, USA, the legs moved apart in a lateral direction as the arch connecting them 'flattened' and the key stone settled by 50 mm. The structure was strengthened to restrain further movement by fitting four tie-bars of 25 mm diameter into holes of 63 mm diameter, and tensioned. When the ties were grouted, the pressurised cementitious grout leaked out of the mortar joints and it was necessary to stop grouting and substitute polyurethane foam. Leakage of pressurised grout through joints in masonry is not unusual and in retrospect it is evident that preventative measures such as repointing beforehand or use of a grout-sock would have prevented it. The refurbished structure is shown in Fig. 10.13.

Lattice

Lattice towers, whether iron or steel, require the usual maintenance of cleaning and painting in regular cycles, depending on the aggressiveness of the local environment. Steel is more susceptible to corrosion than wrought iron and requires more attention as sections can quickly become seriously corroded and weakened. Lattice towers were a popular form of construction in the 1920s and 1930s, and there are many surviving examples, see for example Fig. 10.14.

Where iron or steel sections have become structurally weakened, repairs are sometimes made by bolted cover plates. However, it is difficult to provide adequate protection by painting and further corrosion can quickly

Fig. 10.13 Tower of Wheeling Bridge, USA. Four transverse tie-bars were fitted and tensioned after the masonry arch had settled by 50 mm

Fig. 10.14 Queens Park Footbridge, Chester 1923. Example of a lattice tower typical of the 1920s and 1930s

develop at the interfaces between new and old material. It is generally preferable to cut-out the weakened section and weld in place new material. At the locations where the legs of towers interface with the ground, detritus can become deposited and vegetation can grow around the metalwork. These locations are particularly prone to corrosion. This has sometimes been tackled by casting a concrete plinth on to the foundation to encase the lower steel work, so that the weakened area is supported and the vulnerable interface is raised above ground level. However, locations where there is a steel-to-concrete interface are vulnerable to corrosion and it remains necessary to inspect and clean them regularly. Lattice towers that have become particularly badly corroded have sometimes been encased in concrete up to the saddles carrying the main cables at the tops, for example Alum Chine Footbridge in Bournemouth. This fundamentally alters the appearance and is a solution which may be fully justified on economic grounds, but the preferred treatment of a heritage structure is to replace the weakened parts and retain its appearance.

Iron and steel

Hollow cast iron towers, constructed in the early to mid 1800s, have lasted well and there are several that are still in good condition despite receiving only minimal maintenance. An example of a cast iron tower is shown in Fig. 10.15. Towers having inadequate stability have been successfully strengthened by pre-compression using Macalloy bars located inside the section, as for Caledon Bridge in Northern Ireland.

A similar problem, on a much larger scale, was encountered with the Severn Suspension Bridge when a structural assessment using updated loading requirements showed the towers to be understrength. The solution, to strengthen from inside the towers, was possible as they were stiffened plated structures having sufficient space inside. The strengthened towers were therefore unchanged in external appearance.

Timber

Some of the early footbridges were constructed with timber towers, but have subsequently had more durable materials substituted or been demolished. One of the surviving examples, Burn Footbridge, is shown in Fig. 10.5.

Concrete

Concrete towers were introduced in the late 1800s as riveted iron frames encased in concrete, see Fig. 10.16. This form of construction has performed well, but can deteriorate due to corrosion of the frame causing cracking and spalling of the concrete and it is necessary to monitor conditions for corrosion activity. Steel remained the preferred choice for many years and con-

Fig. 10.15 Stowell Footbridge. An example of cast iron towers

Fig. 10.16 Teddington Lock Footbridge. An example of an early concrete tower

ventional reinforced concrete did not become fully established until the 1960s.

Examples of renovation work and strengthening of towers are summarised in Table 10.2.

CONSERVATION OF CATENARY CHAINS AND CABLES

In the early suspension bridges, both chains and wire cables were used for the catenary (main) cables according to the confidence and experience of the engineer. Samuel Brown was a proponent of chains and used them for the Union Bridge, which was constructed in 1820. Galashields footbridge was constructed in 1816 with wire cables in a cable-stayed configuration and Telford investigated use of wire cables before deciding to use chains for the Menai suspension bridge, constructed in 1826 (see Paxton, 1999). As mentioned earlier, Albert Bridge across the Thames was originally designed to be cable-stayed using wire cables, but they were replaced by chains after the appearance of corrosion. At the time it was argued that the action was premature, possibly unnecessary. Wheeling Bridge, USA, constructed in 1849, was one of the first major suspension bridges to have wire cables.

Table 10.2. Summary of methods of conservation of towers

Problem	Scheme	Comment
Deteriorated masonry	Deal with problem at source. Replace defective stone.	Same techniques as for buildings. Rarely necessary.
Arch settlement (flattening)	Install and grout in place lateral tie-bars.	Commonly used method of restraining movement of masonry walls. Measures have to be taken to avoid leakage of grout through the mortar joints.
Corroded iron and steelwork	Encase in concrete at ground level.	Carried out on several occasions but corrosion likely to occur at interface, requires maintenance. Significant change in appearance. Economic.
	Bolt in place new sections.	Corrosion likely to occur at interfaces. Economic.
	Cut-out defective material and weld in place new material.	Welding of early iron and steel is difficult and requires specialised knowledge and skills. Causes least change in appearance.
Understrength	Encase in concrete up to saddles.	Economic. Significant change in appearance, e.g. Alum Chine, Bournemouth.
	Pre-compress inside section.	Carried out successfully. No change in appearance, e.g. Caledon, Northern Ireland.
	Add structural steelwork inside the hollow section.	Carried out successfully. No change in appearance. A complex operation, e.g. Severn.
Saddle bearings Seized	Free the bearings, redesign or replace components as necessary.	Has occurred on several occasions. If not attended to the towers may be damaged but can often cope with the enforced movement, e.g. Marlow.

They have undergone refurbishment, but are substantially original and in excellent condition after 150 years.

Chains

Catenary chains were composed of parallel eye bars and links, see Fig. 10.17. There was evident concern about the risks of failure and most bridges had two or three parallel chains on either side of the carriageway usually positioned vertically above one another. Wrought iron is less prone to corrosion than steel and in consequence the performances of chains has been generally good. The main requirement for maintenance is due to wear of the pin-joints in the chain assemblage.

There have been several occasions when chains have been replaced by wire-cables to provide extra strength and one where a wire-cable was fitted in addition to existing twin chains. More recently the wrought iron chains of Marlow Bridge were refurbished using new steel links. The longevity of the chains on Clifton Bridge has been ascribed partly to having been proof-loaded so that compressive residual stress was imposed (a technique practised at that time).

The likelihood of fracture of an element of the chain, leading to collapse of the bridge, has to be considered. This occurred in Silver Bridge across the Ohio River between Point Pleasant and Gallipolis in the USA. After some 40 years' service, the bridge collapsed without warning in December 1967 when loaded by heavy traffic, forty-six people were killed and nine injured. It was found that an eye bar had fractured due to stress corrosion or corrosion fatigue (the investigation was unable to conclude which). As a precautionary measure, a similar bridge nearby was dismantled. It has been suggested that components of the chains of other bridges are less highly stressed and are therefore unlikely to fracture. On the occasions when chains have been inspected in Britain using ultrasonic detection equipment, no cracks have been found.

Fig. 10.17 Whorlton Bridge, 1831. Original wrought iron chains

There have been occasions when chains have been badly corroded in vulnerable locations, for example at anchorages. Here the chains have sometimes been strengthened by adding links.

Cables

As confidence grew in the durability of wire cables, they were adopted for long-span bridges in place of chains. Catenary cables of long-span bridges have generally been fabricated on site into bundles of parallel wires. Individual wires have usually been galvanised to provide protection against corrosion. The bundles are squeezed into a cylindrical shape, sometimes a hexagonal shape has been favoured.

Hangers are attached to the cables by clamps, which also act to keep the bundled wires in place. The lengths between the clamps are coated with red lead paste, wrapped in soft wire, and the assemblage is painted to provide a seal against ingress of water. Alternatively, bituminous material has been used to fill interstices between individual wires and provide an external seal.

The catenary (wire) cables of long span bridges have generally been well protected and have exhibited good durability to date. A 9 m length of the main cable of Tamar Bridge was exposed after 34 years and the wires were found to be in excellent condition. In the USA, this form of cable construction was introduced much earlier, a number of the catenary cables are now 100 to 150 years old, and there have been several cases where inspections have revealed various degrees of corrosion in the outer wires. In some cases there was adequate remnant strength, in others it was considered necessary to repair local fractures by splicing and retensioning the broken wires.

The catenary cables of Wheeling Bridge, West Virginia, USA, are composed of parallel wrought iron wires. In 1983 the original protection was supplemented by neoprene wrapping. Unfortunately moisture entered, either through condensation or at the saddles where protection could not be total, and caused some of the wires to corrode. In the 1990s, the neoprene was removed to enable the wires to be inspected and repaired. Broken or badly corroded wires were repaired by splicing lengths of new wire connected by a ferrule at one end and tensioned by a splice clamp at the other. A traditional wire wrapping was applied to provide protection to the cable. This was selected as having a proven history of good performance over many years but in any case, it is more appropriate for a historic bridge. At one of the anchorage chambers there had been leakage and consequential corrosion causing numerous wire fractures and up to 35% loss in the area of the anchor bars. As it was not possible to repair all the wires, an auxiliary support system having high strength threaded bars was fitted to bypass the corroded sections. A so-called fail-safe device fitted across some of the broken wires in one of the cables outside the anchorage is shown in Fig. 10.18.

The catenary cables of shorter span bridges and footbridges have generally been prefabricated in factory conditions and transported to site. Prefabricated cables were commonly used for hangers and stays. There are several types of configuration as illustrated in Table 10.3.

- *Spiral strand* is the most common prefabricated configuration, having successive layers of wire wound round a central king wire. The inclinations of the helices are alternated for each layer to avoid unwinding during transport and at the curvature over the tower saddles when under load. This configuration has the highest packing density and elastic modulus.

- *Locked coil strands* are wound in a similar manner to spiral strand but have the outer layer, or layers, of non-circular wires which are z-shaped to interlock with each other and leave no interstices where corrosion could occur. In practice, locked coils have not prevented corrosion. This is probably due to the imperceptible opening and closing of gaps between adjacent wires caused by live loading permitting ingress of moisture and contaminants which become trapped.
- *Rope* is composed of a group of strands wound in a helix round a central strand. This configuration has a lower packing density and presents an increased area of wire exposed to corrosion. There are also a larger number of interstices where contaminants and detritus can collect. In practice, very severe corrosion and multiple wire failures can occur in the central strand. Rope has a low and uncertain modulus of elasticity.
- *Parallel wire cable*, spun *in situ*, is the configuration commonly used for catenary cables on long span suspension bridges.

Protection of prefabricated cable has usually been by galvanising of the wires and coating the outer surface of the assemblage with a suitable paint. The going life for spiral strand, protected by galvanising and painting is about 60 years, but much longer when given modest maintenance. When wrapping is applied after erection of the cables, it is difficult to provide adequate protection at anchorages and saddles, as illustrated by Fig. 10.19.

Fig. 10.18 Wheeling Bridge catenary cable. Collar or fail-safe device to bypass corroded wires

Table 10.3. Common configurations of wire cable

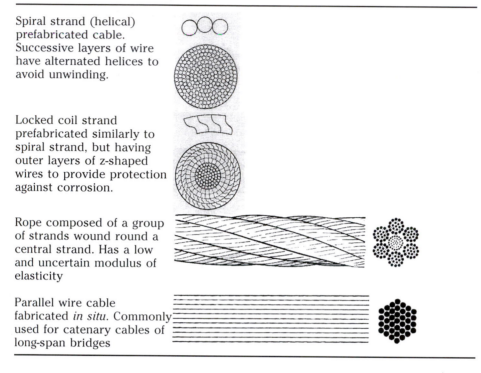

Spiral strand (helical) prefabricated cable. Successive layers of wire have alternated helices to avoid unwinding.	
Locked coil strand prefabricated similarly to spiral strand, but having outer layers of z-shaped wires to provide protection against corrosion.	
Rope composed of a group of strands wound round a central strand. Has a low and uncertain modulus of elasticity	
Parallel wire cable fabricated *in situ*. Commonly used for catenary cables of long-span bridges	

Fig. 10.19 Protection of main cable at saddle. The cover plate has been raised to enable the main cable to be inspected

Fig. 10.20 Main cable having failed protective wrapping. Individual wires have become exposed, but galvanising is providing a second line of protection

Here a cover plate, fitted over the saddle, has been raised to enable the exposed cable to be inspected.

An example of a main cable having failed protection is shown in Fig. 10.20.

Protective tape has sometimes been used for main cables of suspension footbridges, see for example Fig. 10.21. In this example, a metal fence has

Fig. 10.21 Main cable of a suspension footbridge. Damage to protective tape

been built too close to the cable and there has been chaffing and consequential damage to the protection.

Anchorages are often damp and difficult to keep dry. When faced with corrosion damage the space is often filled with concrete so that the cables are fully embedded. However, the interface where the cable enters the concrete must be designed to avoid moisture collecting and requires continual maintenance as it could present a site for future corrosion.

Cable clamps are a common location for corrosion. On occasions when the clamp is badly corroded it may be practical to replace it after carefully cleaning and protecting the cable and hanger. When the cable is corroded, and there is significant loss in strength, it is likely to be necessary to replace the entire cable. This has sometimes been done by fitting the new cable, transferring the hangers from old to new, and removing the old cable. On other occasions it has been necessary to provide temporary support to the bridge, remove the old cable and fit the new one. When the cable is less seriously corroded it is sometimes practical to clean off the corrosion products, provide fresh protection, and clamp a collar over the affected length. The collar acts in the dual role of providing local strengthening and acting as a connector to the hanger.

Puente Duarte Bridge in the Dominican Republic is an example of a long-span suspension bridge where it was necessary to replace the main cables whilst keeping the road open to traffic. The bridge, built in 1955, has a main span of 175 m. When inspected in 1992 it was found, among other things, that the main cables had deteriorated to the extent that it was necessary to replace them. Each cable was composed of twenty spiral strands of 55 mm diameter held in open array by hanger clamps. The galvanising had deteriorated, the lubrication system had mostly disappeared and the paint protection system had failed. The general corrosion of the wires was particularly severe in one anchorage which had become waterlogged. The main cables were replaced by four cables each composed of nineteen galvanised wire strands of 19 mm diameter, compressed into a bundle and clamped together. The protection system comprised modified polymerised linseed oil applied during manufacture, traditional wire wrapping around the bundled strands and a coating of flexible paint to cope with relative movements. The new cables were attached to anchorages positioned behind the originals. New saddles were designed to be fitted to extensions on the towers so that as load was transferred from old to new cables, relative vertical movements could take place.

When a suspension system is found to be seriously understrength, and other methods of strengthening are either impractical or too expensive, as a last resort, it may be necessary to provide additional support by building an intermediate pier or piers, as for Albert Bridge, London.

Examples of renovation and strengthening main cables are summarised in Table 10.4.

CONSERVATION OF HANGERS AND STAYS

Hanger rods

Early hangers and stays were constructed from wrought iron rods which have been generally resistive to corrosion. However, hangers in general

Table 10.4. Summary of methods of conservation of catenary chains and cables

Problem	Scheme	Comment
Worn chain links	Replace with steel.	Carried out on several occasions. Steel has been used but wrought iron is preferable. No change in appearance, e.g. Marlow.
	Replace chain with wire cable.	Carried out on several occasions. Change in appearance. Economic.
Understrength	Add extra cable.	Carried out successfully, e.g. Union Bridge.
	Replace chain with stronger wire cable.	Carried out on several occasions. Change in appearance, e.g. Hutton Ambo footbridge.
	Replace cable system.	May require propping the superstructure. Expensive. No change in appearance, e.g. Alum Chine, Bournemouth.
	Add intermediate pier(s).	Carried out successfully. Significant changes in appearance. Can be more economic than alternatives, e.g. Albert Bridge, London.
Local corrosion	Add chain links at vulnerable location.	Carried out successfully. Economic.
	Splice lengths of wire to repair local damage in wrapped cables.	No reported examples in Britain. Carried out successfully in the USA, e.g. Brooklyn Bridge, New York.
	Clamp a collar over the weakened length.	It is essential to clean the wires and halt further corrosion beneath the clamp. Carried out successfully. Economic, e.g. Wheeling Bridge, USA.
	Fill affected area of anchorage with concrete.	Alters the nature of the anchorage. Need to inspect regularly to avoid corrosion at interface. Carried out successfully. Economic, e.g. Victoria Bridge, Bath.

have experienced fatigue problems, particularly at mid-span of the bridge where they are shortest. This has been due in part, if not entirely, to the failure to recognise the bending action that can be imposed by live loading. Articulation in the longitudinal plane is usually provided in the connections, albeit it is evident that, in practice, few of these joints work, as paint is invariably unbroken across the connection. In consequence, high bending stresses are developed and there have been numerous cases of fractured hangers, for example Marlow (rods), Severn (cables), and various foot-bridges. Most of the fractures have almost certainly been caused by fatigue, but there have been a few where it was reported that the hanger had 'necked' due to corrosion.

In France, a suspension bridge across the Loire at Sully collapsed in 1985 when a lorry collided with a hanger during cold weather when temperatures had fallen to −22°C. It was considered that a steel hanger had fractured in a brittle mode triggering a sequence of fractures and total collapse of the four suspension spans. Although this collapse involved a combination of factors unlikely to occur elsewhere (low ambient temperature, steel believed to have low fracture toughness, and a design susceptible to progressive collapse), it serves as a warning.

Depending on the circumstances, bridges having fractured hangers have been repaired in different ways. Rods have been replaced like-for-like and sometimes the fractured halves have been butt-welded together or connected by a welded splice, see Fig. 10.22. However, if the fractures have been caused by fatigue, a like-for-like replacement is likely to fracture again. This may be sufficient for the requirements of the bridge, otherwise some other

Fig. 10.22 Hanger rod having welded splice repair

action must be taken. It should be added that if a welded connection that has failed under fatigue is repaired like-for-like (by a similar size and type of weld), and continues to be subjected to cyclic loading, it will fail against but in about half the number of cycles of the first failure.

On occasions when it is considered necessary to have an improved performance, stronger hangers have been substituted, usually wire cable in place of wrought iron rod. Stainless steel rods have been substituted on occasions when hangers corroded and failed. However, stainless steel has to be used with care as contact with ordinary steel can develop intermetallic corrosion, see Fig. 10.23.

Hanger cables

For hanger cables, the weakest part is the connection with the socket. When tested to failure some of the early hangers have failed suddenly and in a brittle manner at the sockets. Ultrasonic inspection has been used successfully to detect failed wires, which usually occur in the vicinity of the sockets.

In the case of the Severn Bridge, individual wires were found to be fractured in the vicinity of the socket, but the hangers were not severed. Metallographic examination showed that the mechanism of failure was corrosion fatigue, see Fig. 10.24. The hanger was redesigned to have a larger diameter cable and a connection having transverse articulation, as well as longitudinal, as shown in Fig. 10.25. The transverse articulation was introduced to allow for the racking action caused by the passage of heavy vehicles.

The connection to the socket can also present potential sites for the com-

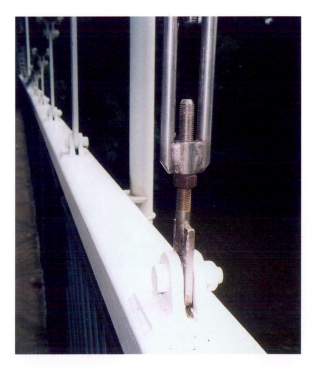

Fig. 10.23 Stainless steel hanger fixed by a mild steel nut. Note the intermetallic corrosion of the nut

Corrosion fatigue crack

Fig. 10.24 Metallographic section through wire removed from a hanger cable. The crack is typical of corrosion fatigue

Fig. 10.25 Improved articulation. Hanger connection having transverse, as well as longitudinal articulation

mon form of corrosion as evidenced by Fig. 10.26. Here the hanger cable has been protected by a polymer sheath which has been inadequately sealed against the socket. The cable has been exposed at the entry point into the socket and serious corrosion has occurred.

Fig. 10.26 Failed corrosion protection. Severe corrosion has occurred at the entry point into the upper socket

As for catenary cables, hanger cables have usually been protected from corrosion by galvanising of the individual wires during manufacture and painting the assemblage after erection. It has the added advantages that the wires are easy to inspect and it is convenient for maintenance work. Recently, there has been a move away from galvanising due to a fear of hydrogen embrittlement but no such failures have been identified to date.

When hanger cables have been observed to exhibit significant vibration due to the effects of wind, vehicular traffic or pedestrians, they can be corrected by fitting Stockbridge 'dampers', see Fig. 10.27. Although they are usually called dampers they are actually small weights fitted at locations on the cable so that the frequencies are changed and resonances are minimised or eliminated under normal use. Alternatively added structural damping can be provided by automobile-type shock absorbers fitted between the cables and deck. This has been done successfully on several cable-stayed bridges.

Stays

Stay-cables, like hangers, can be lively under dynamic excitation. This is usually discovered soon after construction and appropriate action taken, for example by restraining the cables with transverse ties or the provision of additional damping. Where this has been done it is necessary to inspect the system regularly because ties can, and do, fail. It cannot be assumed

Fig. 10.27 Stockbridge dampers. Used to minimise vibration of hanger cables

that an old bridge having a record of stable behaviour will never give problems as rare events can occur which may require attention.

For hybrid bridges having more than one system of cables there may be situations where they can rub together and cause damage. This can be overcome by fitting clamps that prevent abrasion whilst allowing relative movement and correct distribution of loads between hangers and stay-cables, see Fig. 10.28,

Examples of the methods for conservation of hangers and stay-cables are summarised in Table 10.5.

CONSERVATION OF DECKS

Many of the early suspension bridges were damaged by wind and their decks had to be strengthened soon after construction. Others, designed to carry horse-drawn vehicles, have been strengthened successively to carry the increases in traffic loading that have occurred over the years. In consequence, many of the decks of historic bridges are composed of a mixture of new and old materials. Some of the strengthening that had to be carried out soon after construction has become historically significant in its own

Fig. 10.28 Wheeling Bridge, USA. Clamps fitted to prevent abrasive action between adjacent cables

right, for example the introduction of stiffening, whereas other more recent strengthening has been carried out on elements beneath the running surface where it is hidden from the general public. Interestingly, on occasions when strengthening is carried out, the opportunity is sometimes taken to restore parts of the structure to their original form. When assessing maintenance requirements, it is necessary to take into account differing characteristics of new and old materials, for example the lower corrosion resistance of steel in relation to wrought iron.

There are few historic suspension bridges still in original condition that are capable, or potentially capable, of carrying full traffic loading to current requirements. In consequence, most have weight restrictions albeit there is no guarantee they will be respected. Bailliery Bridge in Northern Ireland had a 2 tonne restriction, but was destroyed in 1988 by the passage of a 27 tonne vehicle whose driver had lost his way.

Where strengthening of the deck is considered appropriate and other parts of the structure are adequate, it is necessary to design schemes that are sympathetic to the appearance. Where possible, edge beams are left unchanged and longitudinal beams are fitted inside where they cannot directly be seen.

There are suspension bridges on less significant routes and restricted to pedestrians, that attract less interest and have to be maintained on low budgets. Some are mainly in their original form, but have been around for more than 100 years and are reaching the stage where refurbishment is required. One of the most common requirements is to provide extra lateral stability as original designs tended to rely on the stiffening action of timber decking. This has been done on several occasions by fitting unobtrusive steel cross-bracing.

Table 10.5.   Summary of methods of conservation of hangers and stay-cables

Problem	Scheme	Comment
Fractured hanger rods	Replace like-for-like.	Likely to fail again in a similar time. No changes in appearance. Economic, e.g. Clifton.
	Butt-weld fractured halves.	Likely to fail again in significantly less time. No changes in appearance. Economic.
	Fillet-weld a splice connection between fractured halves.	Likely to fail again in significantly less time. Changes appearance. Economic.
	Replace with stronger hangers.	Life should be enhanced. Appearance likely to be changed, e.g. Trews Weir footbridge.
	Replace with re-designed hangers.	Problem tackled at source. Appearance likely to be changed, e.g. Hammersmith.
Lively hanger cables	Add viscous dampers, friction dampers or Stockbridge 'dampers'.	Viscous dampers require inspection. Little change in appearance, e.g. Severn.
Understrength hangers	Substitute stronger wire cables for rods.	Carried out on many occasions. Changes the appearance. Economic, e.g. Menai.
Corroded hangers	Protect with wrapping.	If normal maintenance is insufficient wrapping may be required. Good detailing required at connection to main cable to keep water out, e.g. Forth Bridge.
Incorrect load distribution between hangers	Check whether loads exceed tolerance.	It may be permissible to accept the position, e.g. Tamar.
	Fit adjustable connections.	Loads can be correctly redistributed. Future maintenance will be easier to carry out.
Lively stay-cables	Fit transverse restraining ties.	Carried out on many occasions. Requires inspection as the ties may fail in the future, e.g. Newport Transporter Bridge.
	Fit additional damping.	Carried out on many occasions. A more elegant solution. Viscous dampers require inspection, e.g. Brottone, France.

Where there is significant corrosion, refurbishment can be carried out by bolting cover plates, cutting-out defective material and welding in place new material, or replacing the entire member. The pros and cons of these methods have been summarised in the section on Towers, see Fig. 10.29 for an example of welded repairs in main girders.

Fig. 10.29 Welded repairs to edge beam of 160-year-old bridge

Many wooden decks have been replaced by more durable materials such as galvanised iron sheet, steel plate and concrete. Where timber has been retained, there is a continuing problem of keeping it waterproof as leakage on to the members beneath causes decay. It is necessary to use good quality treated timber to obtain an acceptable working life and avoid constant maintenance. It is also necessary to have an acceptable skid-free surface.

The decks of suspension footbridges tend to be lively and can exhibit significant vibration. It has been observed that 98% of people walk at pace rates of 1.5 to 2.5 Hz and jog at 2.8 to 3.0 Hz. It follows that decks having first bending frequencies in these ranges are susceptible to excitation. Most of the older bridges have been modified, usually by stiffening, to make them less lively. However, there are some that have survived and are still lively. A by-product of refurbishment can be to bring bridges into the lively range by reduced damping or by making small changes in bending frequency. Lively behaviour and vibration can be a problem if the levels are unacceptable to the general public or the action is sufficient to cause fatigue or structural damage. Exceedance of the static strength of a component, as for the Broughton Bridge, is most unlikely as excessively weak bridges have long since been strengthened or collapsed, albeit there remains exceptions such as Bailliery which cannot be protected against errant motorists. Lively behaviour of decks can be minimised by increased stiffness or added damping. Increased stiffness normally requires the addition of extra steelwork and is a rather inefficient method having the added drawback that it is liable to change the appearance of the structure. Added damping is more elegant and can be designed to be unobtrusive; for example, friction devices can be introduced where there is relative movement at bearings and at joints in the hand rails. Tuned dynamic absorbers (mass-spring-dashpot systems) can be fitted beneath the deck to provide excellent damping, but require monitoring from time to time as the system can drift out of tune. However, it is only necessary to change the dynamic characteristics if there is good reason to do so, for example if the bridge becomes used by a different segment of the public who are less tolerant to vibration, usually the elderly and infirm. Lively footbridges are usually (but not always) located in the countryside where they are crossed by outdoor people having an above average tolerance to vibration. Here, the liveliness is a characteristic of the design which should be retained as part of its originality. A lively bridge that has been in use for 100 years or so and has been accepted by its users is unlikely to require changes to its dynamic characteristics. If it is identified that there is a risk of fatigue damage through the cumulative effects of the vibrations, critical components should be inspected for cracks.

Fatigue, which can be caused by repeated loads or vibration, can occur in welded connections and could become a problem where welding has been carried out in refurbishment schemes. It follows that welded joints require classification and inspection. If a situation is identified where fatigue could cause fracture of a weld and significant loss of strength in a time short of the required life, it will be necessary to take action. There are various options: dress the weld geometry, shot peen to introduce residual compression, remove the weld and substitute a bolted connection.

Examples of methods of conservation of decks are summarised in Table 10.6.

MANAGEMENT PRACTICE

In managing the older suspension and cable-stayed bridges it is not possible to follow ideal principles on all occasions as there are invariably constraints to be taken into account. It is therefore necessary to tailor conservation to suit the circumstances.

Factors that dictate strategy include:

- volume and weight of traffic
- heritage status
- required operational life and
- available finance.

There are numerous suspension bridges over 100 years old and located in highly populated locations, where they are required to carry high volumes of traffic and it is not acceptable to impose excessively low weight limits. It has been necessary to strengthen them by introducing new material to the extent that they resemble the original structures in appearance only. Good practice requires that the appearance is not significantly changed and original material is retained where possible, particularly components in public view, such as edge beams. For these cases the money required to finance the work can usually be found. For bridges in less populated areas not on main routes, the money is less readily available.

Heritage requirements are broadly to limit changes and retain as much of the original structure as possible.

Table 10.6. Summary of methods of conservation of decks

Problem	Scheme	Comments and examples
Deck under strength	Impose load restriction.	Limits likely to be exceeded. Applies to most suspension highway bridges over 60 years old.
	Fit higher strength components.	Change in appearance can be minimised. Expensive, e.g. Marlow.
	Pedestrianise.	A practical and safe method of keeping an historic bridge in continued use, e.g. Victoria Bridge, Bath.
Insufficient lateral strength	Fit cross-bracing.	Little change in appearance. Economic, e.g. Caledon footbridge, Northern Ireland.
Corrosion	Bolt in place new sections.	Corrosion likely to occur at interfaces.
	Cut-out defective material and weld in place new material.	Causes little change in appearance. Needs care to avoid fatigue in the future. Economic, e.g. Victoria Bridge, Bath.
	Replace entire member.	The most satisfactory option. There is no change in appearance. Expensive, e.g. Marlow Bridge.
Sagging	Investigate cause.	May be due to causes that are not acceptable. May indicate causes that are acceptable, e.g. Tamar Bridge.
Deteriorated timber	Replace defective members.	Necessary to use good quality treated timber. Economic.
	Replace with steel deck.	Changes the original structural concept and appearance. Improves waterproofing and surfacing. Expensive, e.g. St Andrews footbridge, Glasgow, Hammersmith Bridge.

Table 10.6. Continued

Problem	Scheme	Comments and examples
Lively deck	Increase stiffness.	Changes to the dynamic characteristics should not be undertaken without good reason. Additional weight. Changes the appearance. Expensive, e.g. Porthill footbridge, Shrewsbury and many others.
	Additional damping.	Friction devices at movement joints have been used successfully. No change to appearance. Economic, e.g. Clapton-in-Gordano footbridge, Gattonside Suspension Footbridge.
Fatigue of welded connection	Planned maintenance.	Practical if there is no potential for serious damage.
	Raise weld classification by dressing and/or shot peening.	Practical but requires regular inspection.
	Substitute a less fatigue-prone connection.	Bolted connections can provide better performances.

Inspection

Suspended bridges are often located across rivers or gorges so that access for a comprehensive inspection requires expensive equipment. This has been recognised in some cases. For example, Clifton Bridge has a gantry to facilitate inspection of the underside of the superstructure and safety lines fitted to the catenary chains to enable harnesses to be clipped in place, as shown in Figs 5.21 and 6.17.

The techniques for inspecting the constituent materials are mentioned in the chapters on masonry, iron and steel, timber and concrete.

Components of suspended bridges can experience relatively high cyclic stresses with the consequential likelihood of fatigue damage. The collapse of Silver Bridge across the Ohio River in the USA in 1967, mentioned earlier, is a forceful reminder of the need to be aware of fatigue. The fact that a bridge has stood for 100 years or more without problems, and is subject to a weight limit, is no guarantee that it will continue to do so, particularly if it is subjected to the increased volumes of modern traffic. Where fatigue is suspected, the locations identified as being susceptible should be investigated. These locations can be monitored to determine the severity of the stress ranges, or they can be inspected to determine whether cracking has developed. Monitoring can be carried out using strain gauges to measure the stress ranges imposed by known traffic loading. The fatigue life can then be calculated using standard methods. To carry out an effective inspection for fatigue it is essential to be within touching distance. There are various methods of detecting cracks, for example, ultrasonic and magnetic particle detection. It is usual to concentrate on the locations where cracking is most likely to initiate. These are at geometric stress concentrations and defects. In some circumstances, corrosion can generate pitting sufficient to cause fatigue cracking. Although it is not generally recognised, stress concentrations are also caused by raised details such as can be present on castings having the manufacturer's name incorporated.

The individual wires in catenary cables, hanger cables and stay cables can fail in fatigue, usually at the point of entry into sockets. The cracking can initiate at surface defects, corrosion pits or it can be due to relative movement between adjacent wires causing fretting fatigue. In cables, fatigue fractures usually occur in outer wires and can often be detected directly, or indirectly, from the rumpled appearance of the cable. Fracture of inner wires can be detected using electro-magnetic equipment.

There is a risk of fatigue in locations where welded repairs have been carried out, cracking can initiate at the weld toe or the root, depending on the regime of stress. As for chains, fatigue cracks can be detected manually, by ultrasonics or magnetic particle detection.

Wire cables require to be inspected for corrosion as contaminants can become lodged in the interstices between adjacent wires anywhere along their length and provide corrosive conditions. Corrosion is most likely to occur at connections between hangers and catenary cables, and at the sockets of hanger cables.

When cables have been protected by tape or wrapping, locations susceptible to corrosion are at the cradles where the wrapping is often stopped, see Fig. 10.19, and at sockets as shown in Fig. 10.26.

Hanger rods and cables should be inspected to check that loads are correctly distributed. This can be carried out in two stages. The first stage is by manual impact-echo to identify any hangers that are carrying negligible loads or no load. In this case the impact will simply generate a dull thud. The second stage is more sophisticated and involves fitting accelerometers to enable the Eigen number and frequency to be measured when the hanger is excited, preferably in a controlled manner so that the response is confined to the first bending mode. An alternative and less satisfactory method is to derive the frequency from analysis of the vibrations produced under normal traffic. If the physical properties, including the degree of end-fixity, are known with sufficient accuracy, the load carried by the hanger can be calculated and comparison made with the load that should be carried. In cases where overloads are discovered, it will be necessary to inspect the hanger for evidence of any damage.

Methods of inspection for problems specific to suspended bridges are summarised in Table 10.7.

Maintenance

The most common maintenance activities are concerned with:

- deteriorated masonry (failed mortar, deteriorated stonework, cracking)

Table 10.7.   Methods of inspection

Item	Methods and comments
Access	Access to chains, cables, hangers and stays of longer span bridges can be difficult and expensive. Gantries can be fitted to facilitate inspection of the underside of the superstructure. Safety lines can be fitted to catenary chains.
Fatigue of chains	The common methods of inspection are by manual means, ultrasonics and magnetic particle detection. Most likely locations for cracks are at geometric discontinuitites and surface defects.
Fatigue of rods	The common methods of inspection are as above. Most likely locations are in the short hanger rods at mid-span of catenary suspension bridges.
Fatigue of cables	The common methods of inspection are manual and by electro-magnetic equipment. Most likely locations are at the point of entry into the sockets of the short hanger cables at mid-span of catenary suspension bridges. Catenary cables are less likely to experience fatigue.
Corrosion of wire cables	Manual methods of inspection will usually suffice. It may sometimes be necessary to remove protective tape or wrapping. Most likely locations for corrosion are at connections between hanger and socket, hanger and catenary cable, at cradles and anchorages.
Distribution of loads in hanger cables	Impact-echo testing can be used to detect unloaded cables. Measurement of frequency response enables load to be calculated, but accuracy is dependent on identifying correctly the degree of end fixity.

- rotted timber
- wear
- corrosion
- wet and corroded anchorages
- excessively lively behaviour
- overloaded hangers
- fatigue
- seized bearings

The maintenance of deteriorated masonry and rotted timber are discussed in Chapters 7 and 8.

Table 10.8. Some examples of catenary suspension bridges

Bridge	Date	Engineer	Main Span (m)	Comments
Union, Horncliffe	1820	Samuel Brown	137	Chains, asymmetric configuration with masonry tower at one end and rock anchors at the other. Retrospectively strengthened by addition of main cables, 1903. Two-tonne weight limit.
Menai	1826	Telford	177	Chains (hanger cables substituted for rods and deck reconstructed c1940), masonry towers. Renovated 1989, no weight limit.
Marlow	1831	Tierney Clarke	69	Chains, masonry towers. Renovated and strengthened 1966. Five-tonne weight limit.
Victoria, Bath	1836	Dredge	46	Chains, masonry towers. Renovated 1940s and 1995. Now limited to pedestrians.
Clifton	1831 1864	Designed by Brunel Completed by Hawkshaw and Barlow	214	Chains and masonry tower. Anchorages partly infilled with concrete. Mostly original. Four-tonne weight limit.
Hammersmith	1887	Bazalgette	129	Chains, iron portal-towers. Renovated 1977. Repaired after bomb damage 2000.
Chelsea	1937	Feneday	104	Articulated steel towers, stiffened plate girder.
Tamar	1959	Anderson	335	Cables, reinforced concrete portal-towers and stiffened truss deck. Strengthened 1990s.
Forth (road)	1964	Roberts	1006	Cables, steel deck with stiffener truss. Strengthened 1990s.
Severn	1966	Roberts	988	Cables, inclined hangers, aerodynamic box deck. Strengthened 1991.

Table 10.9.   Some examples of cable-stayed and hybrid bridges

Bridge	Date	Engineer	Main Span (m)	Comments
Crathie	1834	Justice	42	Hybrid cable system. Downgraded to pedestrian use in 1857. Substantially renewed in 1885.
Albion, Twerton,	1837	Motley	37	Twin towers, harp cables (demolished 1879).
Albert, London	1873	Ordish	137	Hybrid. Strengthened 1880s. Two piers added 1973 to prop centre span.
Newport Transporter, Gwent	1906	Arnodin	197	Hybrid. Stays replaced in 1970s. Closed 1985. Renovated and re-opened 1995.
George St Newport, Gwent	1964	Charles Brown	152	Twin towers, harp-stays.
Wye, First Severn Crossing	1966	Roberts	235	Two single towers, single cables. Strengthened and cables reconfigured in 1991.
Erskine, Scotland	1971	Kerensky	305	Two single towers, single cables. Strengthened 1980s.
Lyne, Surrey	1979	Kretsis	55–55	Twin towers, harp/fan cables 27° skew concrete railway bridge. Said to be the ugliest bridge in Britain.

Wear can occur in moving parts such as the pin joints in chains. This has been discussed briefly in the section on chains.

Corrosion is a common problem that can affect all metal components, particularly cables. The locations on cables that most commonly corrode are the hanger connections, saddles and anchorages. Corrosion of steelwork can occur at ground level and locations, such as re-entrant geometry where detritus can build up. Corrosion can be minimised by regular maintenance cleaning and painting. There are some good examples of well-maintained bridges having cables still in original condition after being in service for 70 years or more.

Anchorages for catenary cables are commonly below ground level where they can become partially filled, or entirely filled, with silt and detritus. This retains water and de-icing salts to create a corrosive environment. The anchorage chamber should therefore be regularly cleaned and drainage channels kept open. The cables or chains should likewise be inspected for

Table 10.10.   Some examples of suspended footbridges

Bridge	Date	Engineer	Main Span (m)	Comments
Galashiels	1816	Lees	34	Cable-stayed having overlapped fans. Destroyed by floods in 1829.
Kirkton of Glenisia	1824	Justice	19	Cable-stayed having harp stays. 16 mm diameter wrought iron stay rods. Still in use and in original condition.
Gattonside	1826	John Smith	91	Conventional suspension configuration having twin chains on each side and hanger rods. Rehabilitated in 1991–1992, sadly losing most of its original ironwork.
Stowell Park	c1840	Dredge	22	Privately owned and closed to the public. The ironwork is in original condition but the timber deck has become unsafe.
Backswood	c1870	Anon	16–10	Two-span conventional suspension configuration. A typical simplistic footbridge having steel towers and wire rope cables. Many unconventional features. Very lively.
Teddington Lock	1889	Pooley	49	Conventional suspension configuration. Iron frame towers having their legs encased in concrete and cross-beams exposed. Twin-cables (wire ropes) on each side and hanger rods. In original condition having required no significant refurbishment.
Queens Park, Chester	1923	Rowell	83	Conventional suspension configuration. Steel lattice towers. Twin-cables (wire ropes) and hangar rods. Typical of the generation of suspension footbridges built in the 1920s and 1930s.
Burgate, Hampshire	1948	Gifford, HCC	30	Steel portals, wire ropes and steel hanger rods. Steel trough deck filled with lightweight concrete. A unique footbridge in original condition.

corrosion. Protective paint or other corrosion protection should be applied. In cases where the anchorage has been filled with concrete, the interface between concrete and the chain or cable requires regular inspection and maintenance as it is vulnerable to corrosion.

Lively behaviour is mainly restricted to footbridges as the older highway bridges have mostly been stiffened in their early years. Vibrations can be caused by wind and by pedestrians walking or running at rates that coincide with bending frequencies in the structure. Lively behaviour as such is an irritant to pedestrians and can be troublesome to elderly and infirm people. It is a challenge to vandals which could provoke them to vibrate the bridge sufficiently to cause damage. It follows that it is prudent to take maintenance action to ensure that the inherent values of damping are retained. Methods of dealing with excessively lively behaviour are outlined in the section on hanger cables.

Fatigue is fairly common in short hangers around mid-span. It is usually caused by longitudinal and transverse bending action of traffic, pedestrians or wind loading. It may suffice to replace fatigue fractured hangers on a like-for-like basis if the life to fracture has been sufficiently long. When fatigue lives are unacceptably short, it is necessary to consider redesigning the hangers to introduce adequate articulation. Other methods are discussed in the section on hanger cables.

In cases where there is an unacceptable distribution of load between the hanger cables it will be necessary to take appropriate action. Redistribution of load may be difficult when there are no means of adjusting the hangers and for these cases it may be necessary fit special connectors (hanger cables were usually made to size and there was no way to adjust them during construction or subsequently).

It is not uncommon for bearings in the suspension system to be seized and inoperative. These are at the tops of towers where the catenary cables should slide over cradles, and at anchorages of hanger cables where there are usually pin joints to permit rotation (as described earlier). Ideally these bearings should be lubricated and kept free to move, but this is usually not possible and in any case many have never been capable of articulation from day one. Fortunately, the structures have usually been able to cope by alternative movements. The exception is the short hangers at centre-span of catenary suspension bridges, where the enforced bending can lead to fatigue fractures.

The above has been concerned almost exclusively with early suspension bridges and early cable-stayed bridges. The more recent cable-stayed bridges are relatively young by comparison and have not yet developed a track record. Nevertheless there have been several where the stay-cables exhibited excessively lively behaviour to the extent that it was necessary to install extra damping and restraint. The more recent forms of construction bring different maintenance issues, for example:

- fatigue of welded connections in orthotropic steel decks
- corrosion-fatigue of wire cables and
- corrosion of reinforced concrete.

Fortunately these issues have been well researched in recent years and there have been numerous advisory documents published (see Gurney, 1992).

Suspension bridges are generally regarded as being rather prestigious and tend to be managed with more care and respect than the more common types of bridge. In consequence, there are more examples of good practice than bad practice. Examples of the latter that have been observed include:

- chaffing of wrapping protection to main cable due to a metal fence being erected too close
- concreted anchorage having poorly detailed design likely to lead to corrosion at the concrete interface
- welded splice repairs to broken hanger rod increasing susceptibility to fatigue
- contact between stainless steel hanger rod and mild steel nut, leading to intermetallic corrosion of the nut
- inadequate connection between protective sheathing and sockets leading to exposure of hanger cable and socket and
- unnecessary replacement of historic ironwork.

BIBLIOGRAPHY

The Institution of Civil Engineers, various proceedings.

Civil Engineering Heritage Series (1981 to 1998), Publ. Institution of Civil Engineers, London.

Ruddock, T. (1999) *Blacksmith Bridges in Scotland and Ireland, 1816–1834*. Proceedings of International Conference on Historic Bridges to celebrate the 150th anniversary of the Wheeling Suspension Bridge. 21–23 October 1999, West Virginia University Press.

Paxton, R.A. (1999) *Early Development of the Long Span Suspension Bridges in Britain 1810–1840*. Proceedings of International Conference on Historic Bridges to celebrate the 150th anniversary of the Wheeling Suspension Bridge. 21–23 October 1999, West Virginia University Press.

Pugsley, S.A. (1968) *The Theory of Suspension Bridges*, Edward Arnold, London.

Gurney, T.R. (1992) *Fatigue of Steel Bridge Decks.* TRL State-of-the-Art-Review, HMSO.

Chapter 11

Movable bridges

GLOSSARY

Rack	A bar having teeth or indentations which engage with teeth of a wheel, pinion or worm, for the conversion of circular motion into linear motion. The rack may be linear or curved, and may be attached to either the moving elements of the bridge or the supporting structure.
Pinion	Spindle or axle having cogs or teeth which engage with the teeth of a wheel or rack.
Pintle	A pin or bolt on which some other part turns, e.g. the main support to a swing bridge, and about which the bridge rotates. The pintle provides lateral restraint to the bridge and may be the principal support or may act in conjunction with other supports such as rollers.
Turnbuckle	In-line component to allow the fine adjustment of tension of ropes and cables, usually employing threaded rod.
Capstan	A wheel and axle arranged on a vertical axis. Pushed around manually by poles inserted into the head of the capstan. Used directly to wind cable or rope on to drum, or through gears to operate other devices.
Windlass	A wheel and axle arranged on a horizontal axis used to wind in cable or through gearing to operate other devices.

Gudgeon	A pivot, usually of metal, fixed to the end of a beam, and on which something turns. The gudgeon pin is the point of attachment of the actuating arm to the structure of a bascule bridge and is normally located at the bridge centroid. The gudgeon post supports the gudgeon pin and provides the connection to the main structural members.
Trunnion	Each of a pair of opposite 'gudgeons' on the sides of a bridge (originally on a cannon) and about which the bridge pivots.

BACKGROUND

Movable bridges have played an important role in the development of bridge engineering and this has been recognised by the number that have either been listed or recorded as having historic significance. Numbers are diminishing as uses of inland waterways change and the movable bridges are either demolished or fixed in position. It follows that there is a need to carry out conservation work on some of the more significant survivors. Information is rather sparse as it is limited to the more notable examples and surprisingly little has been written about movable bridges *per se*. This chapter, therefore, deals with a gap in the technical literature that has not previously been fully addressed.

The best-known movable bridge in Britain, Tower Bridge, London, is a good example of proactive conservation as it is both historic and has to be maintained as one of the major crossings of the River Thames. Over recent years, considerable resource has been expended in maintenance and conservation, and in developing the bridge as a significant London tourist attraction in its own right. Yet the primary purpose of the bridge remains to allow shipping to move up the Thames, whilst minimising delay to the significant volume of road traffic and pedestrian users of the bridge. To this end modern power plant and control systems have been put in place, whilst preserving the appearance of the bridge.

Engineers have devised numerous systems for lifting, dropping, folding, rotating and retracting a span to provide temporary clearance for shipping. The various bridge types can be categorised by movements corresponding to four of the six degrees of freedom.

- Bascules and drawbridges rotate vertically
- Swingbridges rotate in plan
- Lift bridges operate vertically and
- Retractable and transporter bridges operate along the line of the bridge.

The Gateshead Millennium Footbridge across the River Tyne extends the range to a fifth degree of freedom by rotating about the axis of the bridge.

While examples of all bridge types can be found crossing the smallest of spans (usually canals), there exists a progression from drawbridge to swingbridge to bascule bridge to lift bridge, which can be related not only to the size of the span crossed, but also to technical advances in design and materials and to levels of economic activity. Indeed the individual histories of many of the movable bridges of today reflect this progression.

The majority of movable bridges in Britain were constructed in the late 1800s and early 1900s giving access to inland ports. This corresponded to a period of growth of trade by shipping. At the time the volumes of road traffic were modest by today's standards and a movable bridge provided an economic solution, being able to be constructed at river level. As both shipping and traffic levels increased, the delays to road traffic in particular became more significant, and it became economic to consider high level crossings or alternative routes.

During the later part of the twentieth century, the amount of inland shipping has declined and many movable bridges have been replaced with fixed bridges, or have been fixed permanently in the closed position. In consequence movable bridges have generally become less important in Britain. Where they have survived, the older ones are now becoming historic and require conservation. To date most of the work carried out on movable bridges has been in the nature of rehabilitation, where the bridge is treated as having little intrinsic value but worth keeping. This is exemplified by timber canal swing bridges which have sometimes been replaced by similar steel structures. Conservation has generally been limited to outstanding structures, such as Tower Bridge, London, Tyne Swingbridge and the Transporter Bridges at Middlesborough and Newport, Gwent.

Performance

The performances of movable bridges in relation to serviceability have been rather variable, depending on the type of movement, materials and construction. Moreover, due to their moving parts, they require more maintenance than fixed bridges constructed in similar materials. Many of the early movable bridges, particularly drawbridges and swingbridges, were constructed in timber and the maintenance costs of replacing decayed members led to modification and reconstruction in more durable materials. In this sense, the early movable bridges constructed in cast iron and wrought iron have performed much better.

Early machinery was cumbersome and has usually had to be updated to meet modern traffic requirements of speed of opening and closing the highway. Moreover, some movements were steam-driven and outdated. In consequence, machinery has often been partly or wholly replaced by modern equipment and power supplies, giving enhanced performances. This is regrettable as early mechanical and electrical equipment has historic importance in its own right. Where it is still in place, its merit should be assessed and important examples retained.

Damage caused by ship impact is fairly common and some movable bridges have been virtually destroyed to the extent that they have had to be reconstructed or replaced altogether.

In consequence of these factors, few movable bridges have survived to old age in anything like their original form so that, unlike their fixed counter-parts, their long-term performances have not been fully tested. There are exceptions such as the transporter bridges at Newport Gwent and Middles-borough, but these do not have to serve heavily trafficked highways.

TYPES OF MOVABLE BRIDGE

One of the earliest forms of movable bridge was the castle drawbridge across a moat that allowed the occupants to control access to the castle. Medieval drawbridges were relatively sophisticated structures (Fig. 11.1) and readily identifiable as the forerunner of the modern drawbridge. In more peaceful times, most movable bridges are associated with crossings of water where it is impractical or uneconomical to build fixed bridges of sufficient height to permit the passage of vessels. In Britain there are few historic drawbridges. Most examples can be found on canals, to which they remain ideally suited. There are many examples in other countries for example Holland, also associated with canal networks, see Fig. 11.2.

A development of the drawbridge is the Strauss Bridge, sometimes known as a 'jack-knife' bridge as illustrated in Fig. 11.3. In a conventional draw-bridge the size and movement of the balance arm tends to be similar to that of the deck. As spans increase, the tower heights and other dimensions increase in proportion. The Strauss drawbridge incorporates an ingenious, but rather inelegant, counter-balancing mechanism that reduces the oper-ational envelope.

Recent years have seen a number of new drawbridges being constructed, taking advantage of modern materials in port and river estuary situations to cater for leisure craft, rather than commercial shipping.

Through the 1800s, whilst materials improved, the practicalities of con-struction, craneage etc, limited the possibilities for the development of

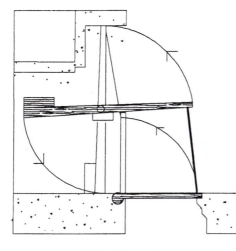

Side Elevation

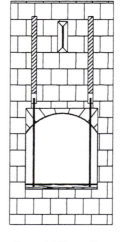

Front Elevation

Fig. 11.1 Castle drawbridge. Method of raising medieval drawbridge

Fig. 11.2 Drawbridge, Amsterdam, constructed 1840

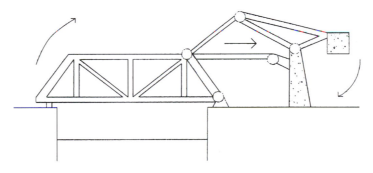

Fig. 11.3 Strauss drawbridge

drawbridges to cater for larger spans. Swingbridges were not so restricted and were widely used in the development of ports and new river crossings. The Newcastle Swingbridge (1876) shown in Fig. 11.4 is perhaps the best-known example from this period.

As ports developed, swingbridges were found to be too slow to operate, sterilised valuable dock frontage and were vulnerable to ship impact. This led to the development of bascule bridges, and particularly rolling bascules based on American designs, which became prevalent in British ports during the first half of the twentieth century. The different types of bascule bridge are shown in Fig. 11.5.

The need for further development of movable bridges declined in Britain, but elsewhere in the world, notably in America, the demand for 'reasonably free, easy and unobstructed navigation' led to the development of the modern vertical lift bridge, see Fig. 11.6, the first being constructed across the Chicago River in 1895. In Britain there is a limited need to operate large

Fig. 11.4   Swingbridge,
Newcastle

Trunnion   Rolling   Rall

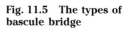

Fig. 11.5   The types of
bascule bridge

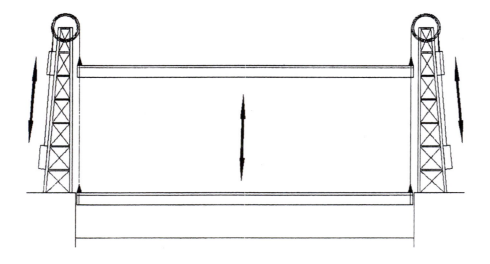

Fig. 11.6   Vertical lift
bridge

vessels inland, and the only significant British vertical lift bridge was the Kingsferry Bridge, which serves both rail and road traffic across the River Swale to the Isle of Sheppey. Kingsferry was constructed in 1960 as a replacement for a Scherzer Bascule of 1904. A lift bridge constructed at New-port, Middlesborough in 1934 was operational until 1990 when it was fixed in its lowered position.

Retractable or telescopic bridges, as the name implies, have a movable span that can be retracted in a horizontal direction, along the line of the bridge, see Fig. 11.7. It 'telescopes' into the existing superstructure. An early example was a footbridge in St Katharine Dock, London, built by Telford in 1829, shown in Fig. 11.28, but no longer in use. One of the remaining examples still in use is a footbridge in Shoreham Harbour, West Sussex, having concrete approach spans and a movable span composed of a steel truss.

Some examples of the different types of movable bridge are given in Table 11.6 at the end of this chapter.

COMPONENTS OF MOVABLE BRIDGES

Foundations

In considering the adequacy of the foundations as part of an assessment or renovation project, the principles are no different to those considered for a fixed span, but particular attention should be paid to the following effects:

- Self-weight of the movable structure is likely to be the predominant load effect, particularly in supporting swingbridges or bascules, where the full structural weight may be carried through a single foundation.
- The point of application of self-weight on to the foundation may change as the bridge operates, as it does with a rolling bascule. The foundation may therefore be subject to a cyclic loading of considerable magnitude.
- A good understanding of the foundation loading from a movable bridge is essential, as any movement or settlement of foundations may impact upon the operation of the bridge, reducing tolerances between movable components.

Substructures

Movable bridges have a high likelihood of being struck by errant ships and there are numerous examples of collisions causing serious damage. Protection of the substructures, by fenders or other means, is therefore essential.

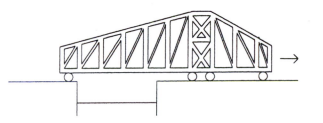

Fig. 11.7 Retractable bridge

This has presented designers with a dilemma in relation to the appearance of both new structures and in the enhancement of old ones. Vessel collision scenarios may be treated as 'normal' or 'abnormal' collisions.

Normal collisions correspond to loading conditions which the bridge (and any protection) should be able to sustain without significant damage. They may include glancing blows by vessels navigating under control and at speeds appropriate to the navigation channel.

Abnormal collisions correspond to more severe loading, which may arise through failure of the vessel's steering, propulsion or tow line, human error, excessive vessel height or extreme environmental conditions. In this case, the vessels may impact the bridge at relatively high speeds and obtuse angles and it may be acceptable for the bridge or its protection to sustain significant damage providing the bridge does not collapse.

In the case of bridges where the consequences of structural failure could be severe, it is likely that a formal risk assessment would be required to define the normal and abnormal collision scenarios and the corresponding loading conditions for design.

The collision risk assessment, together with the preliminary appraisal of operational and physical measures which could be used to control the impact, will enable definition of the impact design criteria for the collision protection systems and definition of the remaining impact design criteria for the bridge itself.

Physical collision protection systems may be installed upstream and downstream of the bridge and/or fixed to the bridge piers themselves.

For normal impacts, conventional fendering similar to that used for ports, docks and locks may be used to absorb glancing collisions and deflect vessels back into the main channel. Such fendering may comprise:

- Rubbing strakes of timber, rubber and/or plastics, fixed to the bridge piers
- Wheel fenders or panel fendering on the cutwater of the bridge piers
- Piled dolphins with fendering upstream and downstream of the piers, possibly providing funnel-shaped 'lead-in' structures either side of the channel.

If there is a large tidal range, then the height of a fixed fendering system would be correspondingly large. In this case a floating fendering system should be considered, which may be guided by fixed vertical piles. Such fendering systems can be effective for glancing impacts at relatively low velocity. However, they may have disadvantages that would in some circumstances outweigh their benefits:

- blockage to the flow of water
- reducing the navigational channel width
- trapping floating debris
- compromising the appearance of the bridge
- cost of construction and maintenance
- cost of replacement.

Therefore, many bridges have no fendering system, relying instead on the strength and shape of the bridge substructure to deflect impacts and on the skills of navigators.

Conventional fendering systems are unlikely to have adequate capacity to absorb abnormal impacts. In this case, two main options may be considered: no fendering system and the bridge substructure having to resist the maximum design impact force, perhaps with significant damage to the bridge, but without collapse; or a collision protection system that may include sacrificial elements which suffer substantial damage, but which would stop or deflect an errant vessel without collapse of the bridge.

In the first option, the bridge substructure may have the mass, strength and foundation capacity to resist abnormal collision loads, albeit with some damage. If these are insufficient it may be feasible to augment their capacity by:

- installation of additional foundation piles, or underpinning with mini-piles
- installed piled skirts or caissons at each end of the piers
- strengthening the piers.

These solutions would be fairly rigid, relying on the progressive collapse of the vessel's structure to expend the kinetic energy of the vessel, and perhaps putting the vessel and crew at significant risk. The peak force exerted by the vessel on the structure would be high.

The second option aims to reduce the peak force on the bridge piers and the structural damage to the vessel, by expending or deflecting the vessel's kinetic energy. Such protective systems may include:

- Rock islands upstream and downstream of the piers, or around the end of the piers. The vessel is stopped or deflected by grounding on the islands.
- Single piles, piled dolphins or caissons upstream and downstream of the piers. Kinetic energy is expended in the progressive ductile collapse of these elements.
- Pontoons or barges moored upstream and downstream of the piers. Energy is expended in dragging heavy mooring anchors.
- Collapsible shells or other energy-absorbing devices installed at either end of the piers.

The most appropriate solution will be dependent on the type of bridge structure, the nature of the collision scenarios and a range of other issues including navigation, dredging, environmental impact and appearance. In general, it is likely that simple, passive, low-maintenance solutions will be preferred.

Superstructures

The superstructure of a movable bridge is subject to a wider range of loading and loading conditions than the comparable fixed bridge.

A particularly important consideration in both assessing an existing movable structure, or designing repair or strengthening schemes, is the relative stiffness of members, both longitudinally and transversely. Generally, a stiffer structure will achieve a better distribution of loading, will limit deflection and distortion of the structure, and allow the structure to be loaded eccentrically, but is consequently more sensitive to support conditions. The relative importance of the flexibility of the components of a movable structure will vary depending on the way in which the structure is operated.

In a fixed bridge, fatigue is normally a consideration limited to members carrying predominantly live load, but in a movable bridge similar effects can occur, with much larger magnitude, under the influence of self-weight during operation of the movable span.

Ballast

Many of the forms of movable bridge rely on the use of ballast or counterweights to balance the self-weight of the main structure and allow the bridge to be moved by the application of fairly modest forces. In the past ballast may have been used in situations when nowadays it is practical to have a fail safe system so that direct force can be used.

When originally designed, the driving force required to operate the bridge would have been determined in relation to the calculated out-of-balance force, with suitable reserve. However, it is important that this relationship is checked and maintained over time to take account of changes in material properties e.g. frictional resistance effects of settlements, or reduced efficiency of the driving or braking system.

Normally the balance arrangement for a movable structure will seek to provide at least small positive reactions in the 'closed' position. In undertaking strengthening or remedial works to the main superstructure, any changes to the weight or to the distribution of that weight, must be accompanied by an appropriate adjustment to the balancing weight. Even the removal of years of accumulated paint systems can reduce the weight by up to several tonnes, and require the structure to be rebalanced.

Techniques are now available to optimise the balance state of a bridge and hence reduce the likelihood of motors being overloaded. These are discussed further in relation to the different types of movable bridge.

Decks

The deck will normally be the principal load carried by the fixed structure, certainly whilst the bridge is being operated, and often when in use by traffic as well. It is therefore important that the weight of the deck and its distribution on to the superstructure is understood before assessing the structure or designing renovation or conservation measures.

Historically the decks of many bascules and swingbridges have been constructed from multiple layers of timber planking, as this is a relatively lightweight material, readily worked to the required shape and capable of being fixed, to allow the deck to be raised to the vertical. Such decks suffer from ingress of water and consequential decay so that they need to be reconstructed with new timbers from time to time during the life of the structure. Timber decks can be treated to provide a suitable surface for pedestrians, and use of modern waterproofing materials can extend the life of the timber. Use for road traffic requires an appropriate form of treatment, such as specialist epoxy-coated surfacing panels to provide the necessary skid resistance in a form that will remain secure whilst the bridge is raised.

A wider choice of materials is available for the deck construction and surfacing of decks that remain horizontal. Capital and operating costs for the bridge opening mechanism will nevertheless still be related to the weight of the deck, and the choice of a lightweight solution is normally preferable.

Lightweight orthotropic steel decks (sometimes called battle decks) have been used for new construction since the 1950s and for refurbishment of older bridges. However problems can arise requiring eventual refurbishment or replacement due to fatigue in weld joints to the underside of the deck plate. Also, premature cracking can occur in the surfacing over hard spots formed by stiffeners beneath the flexible deck plate.

DRAWBRIDGES

In a basic drawbridge, the deck is pivoted at one end and rotated towards the vertical by pulling on the nose of the bridge or at mid-span, using ropes or cables. A force in excess of the full weight of the deck may be needed to raise the drawbridge depending on the geometry. The majority of the force needed can be provided by counterweights, using a relatively small additional force to control the movement. Nevertheless the supporting cables or chains must have considerable strength and need to be inspected at regular intervals. An example of a typical drawbridge across the Grand Union Canal is given in Fig. 11.8.

In order to reduce the force needed to control the structure, drawbridges have been developed with independent mechanisms for balancing the deck weight. The Dutch canal bridges are typical of such bridges, as shown in Fig. 11.2.

Drawbridges have the advantage that all the operating equipment and deck superstructure can be located above ground, avoiding the need for difficult excavations during construction and easing maintenance.

Drawbridges of more significant size are generally controlled by a rack-and-pinion drive, having the advantage of restraining the deck in both tension and compression. The drive system not only has to position the deck, but also has to overcome friction and 'sticking' forces, control wind forces, and cope with a varying geometry as the bridge operates.

Drawbridges require a reasonably stiff superstructure, particularly transversely, to carry wind forces in the raised position without undue distortion. Also, such bridges are often driven from two sides in normal operation, but may need to be capable of being driven from one side only in the event of a breakdown or the need for maintenance work

Fig. 11.8 Typical canal drawbridge

Conservation

Drawbridges were common on canals, where they were often of timber construction, but have been subject to decay caused by the damp conditions and historically poor maintenance. Consequently timber drawbridges are now becoming rare, often being replaced with steel structures. Many have had their timber decks replaced by steel.

Small drawbridges, as typically found on canals, have historically been hand-operated by pulling on a rope attached to the balance arm. Such drawbridges were usually of timber construction, and needed to be rebalanced with the seasons as the moisture content and hence density of the timber varied. Even if properly balanced, there is a risk that the person using the bridge may not have sufficient strength to control it, which has on one occasion resulted in a fatal accident. With increased leisure use on the canals, many of these bridges are being converted to manual hydraulic or electrical operation in order to make them easier to operate and reduce the risk of accidents.

SWINGBRIDGES

A swingbridge is a structure which is balanced about a single point, either by having two equal spans or one span (balanced by the addition of ballast to the shorter span) which rotates in plan through 90°, see Fig. 11.9. Swingbridges require considerable space to operate, and if located over water can be vulnerable to ship impact. With double leaf swingbridges, alignment at the central joint can be a problem.

Swingbridges are not subject to the same degree of potential eccentricity in the transverse loading, and hence transverse stiffness is less significant unless the self-weight loading itself is likely be unsymmetrical. The longitudinal stiffness however will in part determine the amount by which the bridge must be raised (or the nose supports lowered) to allow it to be rotated, and the sensitivity of the structure to any differential settlement.

Swingbridges will generally have different systems for carrying load depending on whether the bridge is being moved or whether it is carrying traffic. The bridge in its operating position must have sufficient positive reaction to prevent load reversal under any envisaged pattern of live loading, but in order to allow the bridge to move it is necessary to remove or overcome all restraining lateral movement. Generally, this will be done by having positive removable supports rather than relying on systems involving sliding. There are various ways in which this has been achieved, but most commonly they involve raising and lowering supports rather than lifting the deck. With increasing loading from modern traffic it will be important that the supports which restrain the deck also isolate the components of the movable mechanism from traffic loading as far as possible. Such systems may involve mechanisms to slide bearings into positions or pairs of wedges operated by a screw mechanism. In either case it may be necessary to apply

Fig. 11.9 Typical canal swingbridge

a temporary uplift force to the deck to allow the frictional forces to be over-come.

In some cases the weight of the swingbridge is carried through a central pintle, with balance wheels to maintain stability. In others, the pintle may serve only to provide horizontal restraint, with the load being carried through rollers.

An equally diverse range of mechanisms has been employed to move swingbridges once they have been made free to rotate. Rack-and-pinion drive systems may involve a fixed rack with the pinion mounted on the superstructure or a fixed pinion driving the rack on the superstructure, see Fig. 11.10.

Other swingbridges are moved using wire ropes or chains bearing on or fixed to a slewing ring and driven by a windlass, Fig. 11.11, or in some cases by extending hydraulic cylinders acting through a system of pulleys, Fig. 11.12.

Conservation

As with other types of movable bridge, it is important that the bridge is correctly balanced, both to minimise the loading on drive systems under braking, and to ensure the load distribution is maintained as required by the design.

Swingbridges, in their movable mode of operation have an inherent large inertial moment, which can apply excessive loading on drive shafts, pinion gears and ring gear when braking is applied. The total force will be determined by operational requirements for the bridge, but balancing of the braking system will help reduce the risk of overloading individual components.

Fig. 11.10 Rack-and-pinion drive

Fig. 11.11   Windlass

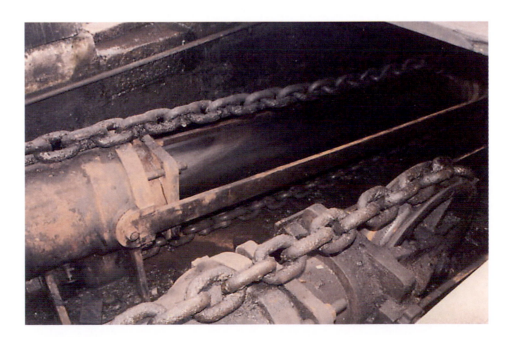

Fig. 11.12   Hydraulically driven slewing chain

Conservation involving refurbishment and strengthening has been under-taken on the grade II listed Town Bridge in Northwich. Constructed in 1899, it has wrought iron and cast iron components. As well as being the first electrically driven swingbridge in Britain, it is unusual by virtue of the bulk of the self-weight of the bridge being supported on hollow pontoon tanks floating within a caisson. In 1998, it became necessary to strengthen the superstructure to meet the 40 tonne traffic requirements, and carry out repairs to the pontoons, which could only be achieved by their removal. The largest crane in Europe was used to lift the superstructure (356 tonnes) and pontoons (180 tonnes), shown in Figs 11.13 and 11.14. The repairs involved a variety of activities.

- The original trough decking was removed and replaced with steel decking, supported on replacement steel plate girders fabricated to match transverse mild steel beams fitted in the 1920s.
- The pontoons were found to be severely corroded over the top 750 mm 'splash zone'. The bridge rotated on seventy-six rollers, which were checked for cracking using gamma rays, and resulted in nineteen being replaced.
- The bridge was operated by a wire slewing rope of 40 mm diameter, driven by a capstan through gearing from an electric motor, and this was restored, but the original 'variable resistor' controls were replaced with modern programmable logic computer systems.
- A sprayed waterproofing system was applied on to the deck plate.
- Closed circuit television and modern traffic management systems were installed.

The project illustrates the compromise which often has to be reached in conservation of historic bridges. The refurbishment has maintained the external appearance and principles of operation, but provides a bridge to current strength and safety requirements.

The modernising of the swing bridge in Albert Dock, Liverpool, is another example of conservation, see Fig. 11.15. Modern stitching techniques were

Fig. 11.13 Refurbishment of Town Bridge, Northwich. Lifting out the superstructure

Fig. 11.14   Refurbishment of Town Bridge, Northwich. Lowering the pontoon (180 tonnes)

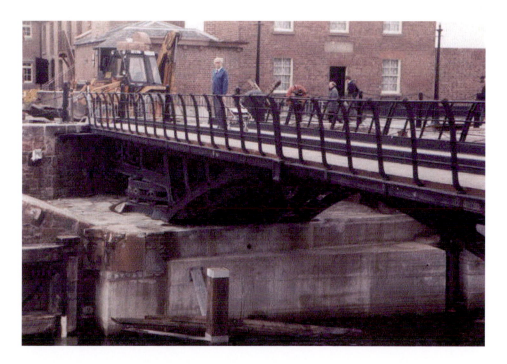

Fig. 11.15   Swingbridge, Albert Dock, Liverpool, 1843. When open to shipping, the parapet is folded down to the deck to avoid being damaged by impact

used to restore the strength of the cast iron ribs to provide emergency vehicle access. The original parapets and hand-operated windlass were retained, but modern deck surfacing replaced the timber block paving. This bridge has an interesting feature where the parapet leans outward from the deck. When opened to shipping it has to be folded back, because if left in position it would be proud of the dock wall and liable to impact damage.

Boothferry Bridge is a seven-span structure built c1930 across the River Ouse in Humberside. It has two swing-spans, having lightweight orthotropic steel decks. The structure was found to require maintenance in the 1970s when, among other things, fatigue cracks were discovered in the welds attaching stiffening ribs to the deck plate. Cantilevered footways on each side of the carriageway were suffering corrosion of the thin steel supporting members and exposed reinforcement in the concrete footway slabs. Corrosion has also occurred in riveted steel supports to the hand rails causing bulging between the rivets. The bridge was repaired and strengthened in 1978. In later work, the old DC machinery was replaced by hydraulic equipment.

Reconstruction

Selby Swingbridge is an example of sympathetic reconstruction. The original bridge was built c1791 by the eminent eighteenth century engineer William Jessop. It was reconstructed in 1921 at the time of a major accident. By 1970 it had deteriorated through a combination of old age, heavy traffic and numerous collisions from shipping. It was the only surviving eighteenth century timber trestle bridge still carrying trunk road traffic. The swing span was operated manually with a windlass and could be opened in one minute. The movement was a notable example of the early use of ball bearings on a large scale. The specification for the reconstruction required that the spirit and concept of Jessop's original design be retained. Where appropriate timber was to be used for the main structural elements and where possible original timber was salvaged and reused. Research was carried out in support of the design: measurements were made of the drag coefficients of the pile bents in flows of 7 to 8 knots, and wheel load tests were carried out to determine the strength of the timber deck panels. The reconstructed bridge is shown in Fig. 11.16.

BASCULE BRIDGES

The word 'bascule' derives from the French word for see-saw. Like a see-saw, bascule bridges rotate and are balanced about a horizontal axis. A relatively small force is therefore needed to move and control the bridge. There are two types of bascule bridge, trunnion and rolling, and there is the Rall bascule, a variant of the rolling bascule, as illustrated in Fig. 11.5.

Trunnion bascule

Trunnion bascule bridges are structures which are balanced about a single fixed point and are rotated vertically to clear the opening, see for example

Fig. 11.16 Selby Swingbridge. Reconstruction in 1970 was sympathetic to Jessop's original bridge of 1791

Fig. 11.17. In order to remain in balance throughout the opening sequence, the trunnion has to be located at the centroid of the leaf, and the counterbalance, which has to rotate below deck level for the bridge to open, requires significant depth within the abutment. Furthermore, it is difficult to achieve a fully vertical position when open, and the span will need to have been designed to allow for the leaf encroaching into the clear opening.

A trunnion bascule may be hydraulically actuated, but was more commonly driven by a rack-and-pinion system. The rack may be attached to the leaf of the bridge with a fixed pinion, or the pinion may be attached to the leaf engaging with a fixed rack. This form of drive provides positive control during both raising and lowering operations.

Span locks hold the lift span in place when it is in the down position. The span locks are released by the operator prior to raising the bridge. Small motors and gears are used to engage the span locks and help align the bridge as it is lowered into its down position. Air buffers (above) soften

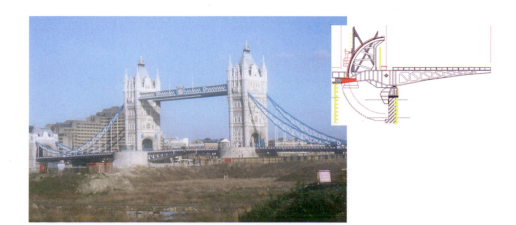

Fig. 11.17 Tower Bridge, 1894. The best-known example of a trunnion bascule

the impact of the bridge leaf as it reaches its maximum limits of opening and closing.

The main motors do not have to be very big, because the weight of the bascule span is balanced by the large counterweights. The gears are connected to the rack assemblies on the bottom of the lift span leaf. The counterweights can be made of concrete and balance the weight of the leaves extending over the river channel. For bridges located in cold climates additional weights can be added to the counterweights to offset any added load on the leaves due to snow or ice accumulation in the winter or construction modifications. The steel structure that makes up each leaf pivots on two trunnions.

Ensuring an optimum state of balance throughout the operating cycle can enhance the reliability and life of machinery driving the bascule. At the most sophisticated level strain gauges can be installed on to driving machinery to measure torque and hence determine load and friction conditions, as well as the state of balance throughout the operating cycle. The balance of the bridge can then be adjusted to achieve optimal balance.

Tower Bridge, London (Fig. 11.17) is an example of a trunnion bascule. In this case the pinion is fixed, just below deck level and drives on to racks at either side of the deck and above it.

The bridge comprises two outer suspension spans supported by chains and the central double-leaf bascule span. The suspension chains are continuous across the central span, being concealed by the high level walkway. Contrary to the general impression, the bridge is essentially a mild steel structure clad in Portland stone.

The bridge was originally operated by water pressure, using steam from two of four boilers pumped into six hydraulic accumulators, which could be released at a pressure of 5.1 N/mm^2 (750 psi) on demand, with a back-up connection to the London Hydraulic Company's main at a similar pressure. The steam was used to drive the bascule engines, of which a total of eight was provided in four engine rooms. Large and small engines catered for different levels of wind loading and could be linked for a wind on the scale of the storm that caused the Tay Bridge disaster, at 0.38 N/mm^2 (56 psi). Each bascule could be driven from either the upstream or downstream side of the span, with the other side available for maintenance or standby. In practice only one engine has ever been used.

Conservation

During the life of Tower Bridge there have been various maintenance requirements. A recurring problem has been water ingress into the surfacing over the bascule deck. Swelling of the original timber infill caused the wood block paving to erupt. Matters were improved by the use of foamed polyurethane replacement infill, but eventually the surfacing was replaced with plywood panels on epoxy mortar with neoprene tiles and epoxy topping.

The major refurbishment works to the movable elements of the bridge have been centred on the replacement of machinery. Operating costs were high due to the number of staff needed and lack of standardisation, requiring most spares to be purpose made. In the mid 1970s the bascule engines were replaced by hydraulic motors driven from a hydraulic power pack with two electric motors and motor starter/control panel. Listed building status required four of the original bascule engines and two pumping engines to be retained *in situ*. At the same time, a hydraulically operated positive support system was introduced to relieve the main pivot shaft bearings of the increasing loads from traffic.

Subsequently, with reducing demand for operation of the bridge, works have concentrated on converting the bridge into a major tourist attraction, involving significant structural repairs and modifications to provide for the circulation of the public. The bridge has a weight restriction and is raised ten to twelve times a month.

Rolling bascules

Rolling bascules were first built by the Scherzer Rolling Bridge Company and are often referred to as 'Scherzer' bascules. Rolling bascules addressed the shortcomings of the trunnion bascule, particularly in dock locations, in not providing adequate clearance at the dock edge and requiring significant abutment works below deck level, see Fig. 11.18.

The deck is connected to the counterweight by a quadrant girder. Actuating arms are attached at the centroid of the structure and pull the structure into the open position. As the bridge opens it rolls on and is guided by the toothed track girders, so that the leaf also moves away from the opening. As with the trunnion bascule, the rack may be attached to the leaf of the bridge with a fixed pinion, or the pinion may be attached to the leaf engaging with a fixed rack.

Poole Harbour Bridge is an example of a double-leaf rolling bascule bridge, as shown in Fig. 11.18. It was constructed in 1927 and strengthened in 1995

Fig. 11.18 Poole Harbour Bridge, 1927. A double-leaf rolling bascule bridge

to carry 37.5 units of HB loading. Other than this the structure has remained in substantially original condition.

Conservation

With a rolling bascule the full weight of the structure, including counterweights, is transferred from the segmental girder along a single line of action into the track girder. As the bridge operates the point of contact moves along the segmental girder and is guided by a series of teeth and holes, which also serve to resist longitudinal forces. Typically the load is transferred from the web of the segmental girder on to the track plate in bearing, but the track plate will be held in place by angles, which also transfer longitudinal shear. During operation of the bridge, flexure of the track plate relative to the webs results in bending at the root of the restraining angles. This effect will be made worse if there is any detritus on the track or any misalignment of teeth.

Circumferential cracking along the connection between the track and web plates is a common defect in such bridges and attempts to repair the angle by welding have proved only partially successful. The track plate can also deteriorate under the effect of the full weight of the bridge being transferred as a highly concentrated line load across the track girder. This causes cracking and wear to rear surfaces and cracks emanating from the corners of holes to locate the teeth in the rolling plate.

Many rolling bascule bridges have had their track plates repaired or replaced one or more times during their life, and these components must be treated as mechanical elements requiring regular maintenance and periodic replacement.

Repairs to rolling bascule bridges usually represent a major undertaking. If carried out *in situ*, the bridge will often have to remain operational, at least to river traffic, but repairs to track girders cannot easily be undertaken on a piecemeal basis. One solution adopted has been to remove the bridge by pontoon to a remote location. This is exemplified by the conservation of Duke Street Bridge, Birkenhead, which is one of a family of rolling bascule bridges designed and constructed by Sir William Arrol around the Liverpool and Birkenhead docks during the 1930s, see Fig. 11.19. General deterioration of the bridge had resulted in heavy corrosion of members and fatigue cracking of the rolling path components in relation to the movable elements. Of particular concern was circumferential cracking affecting the transfer of load between the webs and track plates around the quadrant girder, which was attributed to lack of maintenance generally, but more particularly to ship impact giving rise to mismeshing of the teeth and sockets used to guide the bridge.

It had already been established that the deck would need to be replaced with a shallower form of construction, due to its poor condition, the need to strengthen it to meet current loading requirements, and the need to keep the bottom flange out of the water (whose level had been raised for operational reasons). The operation of bascule bridges requires that the centroid of the movable structure is coincident with the point of attachment of the actuating arm (the gudgeon pin, see Fig. 11.20). Thus changes arising from replacement of the deck would require compensating changes to the ballast.

Fig. 11.19   Rolling bascule bridge, Birkenhead Dock. One of a family of bridges c1930s

Fig. 11.20   Gudgeon pin

It had further been established that the waterway would need to remain open to shipping throughout works, but that it would be impractical either to undertake repairs with the bridge in a vertical position, or to carry out repairs such that the bridge could be opened at short notice. Complete replacement of the structure was considered.

- Construction of a different type of movable bridge was rejected in part because of the planning and procedural delays which would have resulted, but also to retain a sympathy with other bridges in the area.
- Construction to the original design was rejected on economic grounds.

Overall it was deemed more economic to replace both the quadrant girder and ballast box, rather than repair them. This saved 110 tonne of steelwork and allowed the existing track girders and actuating arms to be retained. The bridge was floated by barge to an adjacent dock where repairs could be undertaken on land.

The critical element of this type of movable bridge is the track plate, which has to be accurately fitted around the quadrant girder. Rolling, machining and fitting the bearing plate (750 × 110 mm thick) requires skills and equipment not necessarily available to all steel fabricators.

Extended life may be achieved for a bascule structure by the use of modern monitoring techniques that can be remotely accessed. Continuous monitoring of strain gauges and inclinometers can provide base data during normal operation of the bridge, against which abnormal operation or ongoing deterioration can be monitored. Funds for maintenance can consequently be better targeted and planned with less disruption to users.

The Mitchigan Street Lift Bridge c1930 crosses Sturgeon Bay, Wisconsin, USA. It is a double-leaf rolling bascule, which is opened on average ten times a day. Problems were first identified during routine inspection when movement of the rolling plate was noted where it was attached to the web of the quadrant girder through an angle. Specialist inspection found that all rivets connecting the rolling plate to the connection angle had failed, with only corrosion keeping the rolling plate in place at several locations. Circumferential crackling was also found at the root of the connecting angle and in previous repair welds connecting the angle to both the rolling plate and quadrant girder web. Cracking was also found in the rolling plate emanating from the track teeth sockets.

When the rolling plate was removed for repair, the cracking was found to be more extensive than anticipated. It was not feasible to fabricate new cast steel track plates within the time available for bridge closure, and so extensive weld repairs were undertaken. However, in two locations fully satisfactory repairs could not be achieved and so a strain gauging system was fitted to allow continuous monitoring of the cracks. The system proved successful in that it enabled the continuing operation of the bridge and deterioration of the rolling plate to be monitored.

Rall bascule bridges

An interesting variant of the rolling bascule is the Rall bascule bridge, Fig. 11.21. The design seems to have arisen in part as a way around the patents held by the Scherzer Bridge Company. The rolling track is located at high level, supporting the structure directly at its centroid through the 'Rall' wheels. As the bridge opens it both rolls and pivots on the Rall wheels, with the actuating arm and a lower control arm acting against each other to cause the deck to rotate upwards as it rolls back from the clear opening

Fig. 11.21 Rall wheel. The deck is suspended from this wheel. The system was devised by Theodore Rall to overcome patents

Three Rall bascule bridges are still in existence in the USA. The biggest is Broadway Bridge in Portland Oregon, constructed in 1912. It is a double leaf structure, each leaf being 42.5 m long and weighing around 2,000 tonne. It is not popular with motorists as opening takes 20 min and longer.

VERTICAL LIFT BRIDGES

A vertical lift bridge is a structure supported from towers by cables passing over pulley wheels to counterweights as shown in Fig 11.22. The structure can be raised or lowered by winching against the nominal out-of-balance force. Lift bridges do not require significant additional space outside the span, but can impose a height restriction on vessels using the passage. The deck stiffness for a lift bridge is less significant and unlikely to affect the distribution of loading to the supporting cables, provided the cables are correctly balanced in the first place.

Conservation

In common with other types of movable bridge, the force required to raise the bridge is small compared to the weight of the structure, having to overcome frictional forces and the unbalanced element of the load. If equal tension (and hence extension) is not maintained within the lifting cables, cocking of the span can result during lifting. This in turn results in uneven loading of the mechanical system, increased frictional forces and reduced lives of the mechanical components.

Fig. 11.22 Huddersfield
vertical lift bridge

As with bascule bridges, it is now possible to measure the torque required by the driving system throughout the lift cycle to determine the friction characteristics and subsequently to optimise the state of balance over the full cycle of raising and lowering the superstructure.

A major overhaul and renovation of the famous Duluth lift bridge in Minnesota, USA, Fig. 11.23, took place in 1985–1986. Numerous tasks were undertaken on the bridge, but the major ones included relocation of the heavy lift motors from the centre to new motor houses at the ends of the lift span. Four new 100 horsepower motors were installed, two at each end. Only two are required to lift the bridge, as in its original 1929 design. Electricity for the new motors comes directly from city power lines or a back-up diesel generator. The carrying capacity of the span was increased substantially by moving the lift motors. The work also extended the useful life of the structure.

A new, smaller operators' control house was built in mid-span with state-of-the-art computerised operating systems installed, plus navigation and communication equipment. Lifting cables were replaced, structural members inspected and strengthened where necessary, and corroded rivets were replaced. Additional weight was added to the counterweights to compensate for the increased weight of the centre span from 900 to about 1,000 tonnes. The entire structure was sandblasted and repainted.

One of the problems inherent in the remodelling of the bridge and moving the lift motors to the ends of the movable span, was how to get both ends to rise equally to keep the span level. This was not a problem with the original design, because all the cables wound around a single drum in the middle. However, in the refurbished structure each end operates independently. Movements are managed by a computer that controls the speed of each end both in raising and lowering. If one end gets beyond the tolerance

Fig. 11.23 Duluth lift bridge

allowed by the computer, operation is switched over to manual control. Bridge operators can then fix the problem, level the lift span and continue operation.

The bridge is repainted about every 15 years. So much paint is used that it affects the balance, requiring some changes in the counterweights. The heavy cables are changed every 20 years or as required. Ropes are greased twice a year.

Some components of lift bridges are shown in Figs 11.24 and 11.25.

Regular inspection and maintenance of the ropes is essential to the safe and reliable operation of lift bridges. Where abrasion is an important factor, the rope must be made of a coarse construction containing relatively large wires. In other cases, the severe bending to which the rope is subjected is more damaging. Here, a more flexible construction, containing many relatively small wires, is required. In either case, however, if the rope operates over inadequate size pulley wheels, the repeated bending stresses will cause the wires to break from fatigue, even though actual wear is slight. The smaller the diameter of the pulley wheel, the sooner these fatigue breaks will occur and the rope life becomes shorter.

Another undesirable effect of small pulley wheels is accelerated wear of both the rope and the groove in the pulley. The pressure per unit area of rope on the groove for a given load is inversely proportional to the size of the wheel. In other words, the smaller the pulley wheel the greater the rope pressure per unit area on the groove. Using the correct diameter pulley wheel for the size and construction of rope can prolong both pulley wheel and rope life.

The diameter of the pulley wheel can also influence rope strength. When

Fig. 11.24   Pulley wheel

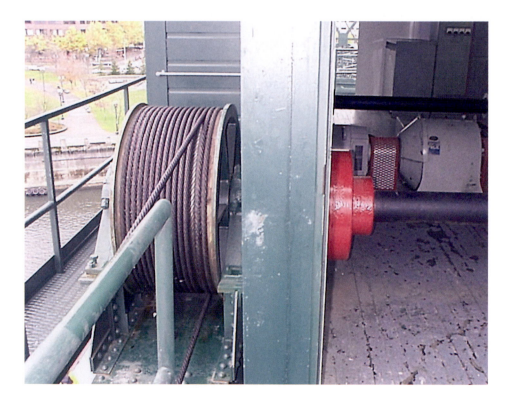

Fig. 11.25   Drive shaft and drum

a wire rope is bent around a pulley wheel, there is a loss of effective strength due to the inability of the individual strands and wires to adjust themselves entirely to their changed position. Tests show that rope efficiency decreases to a marked degree as the wheel diameter is reduced with respect to the diameter of the rope.

Although the sizes of the pulley wheels are determined in the original design, these factors may need to be considered during conservation work if any changes have to be made, for example fitting stronger ropes to carry heavier loads. Also, it may be required to improve serviceability by re-optimising the relative sizes of the pulley wheels and rope.

TRANSPORTER BRIDGES

Transporter bridges have a high level fixed structure above the clearance required for shipping. A carriage arrangement can be pulled across the structure and from this is suspended a gondola which can carry a relatively small number of vehicles and pedestrians. Only four bridges of this type have been built in Britain, of which three remain; at Newport, Gwent, Middlesborough on Teeside and Warrington across the River Mersey. The fourth, at Runcorn, was replaced by the Runcorn–Widnes Road Bridge which opened in 1961.

Newport Transporter Bridge has a 4 m deep boom supported from main suspension cables. The purpose of the boom is to distribute loads uniformly on to the suspension cable. The boom is a pin-jointed arrangement with sets of overlapping diagonal bracing cables. The main suspension cables are anchored at ground level some 137 m back from the towers. Secondary cables anchor the boom. The span between towers (centre to centre) is 196 m.

Middlesborough Transporter Bridge has two deep cantilevered trusses which meet in the middle of the main span. The main cantilevers (85 m each) are balanced by shorter (43 m) end spans which are anchored at their end to the ground by vertical steel cables. The clear span between the towers is 172 m.

Warrington Transporter Bridge is a smaller structure having a clear span of 61 m, Fig. 11.26. It is privately owned and not open to the public.

Fig. 11.26 Warrington Transporter Bridge. The carriage is shown in the inset

There is an operating transporter bridge, 'Die Eisenbahnhochbrücke' at Rendsburg in Germany. This is a high level fixed railway bridge, but with a suspended gondola providing a crossing at canal level for pedestrians and vehicles. The 317 m canal crossing is approached by over 1 km of structure on either side.

Conservation

The transporter bridges at Newport and Middlesborough have received substantial conservation in recent years, and are now operated on a daily basis. Some of the problems encountered are relevant to other structures:

- Original cables, comprising many small (5 mm) wires spirally wound, were found to be of inferior quality to even the most basic grades of wire in use today, but having strengths in excess of that specified at the time.
- Cables removed from the bridges were found to contain both corroded and broken wires and evidence of fatigue damage. The latter was attributed to the sensitivity of fatigue strength to environmental and processing defects.
- The double braced boom structure at Newport proved to be both an advantage and a disadvantage. The resulting high level of structural indeterminacy allowed individual members to be replaced with little need for temporary works, but control of structural stiffness, as affected by retensioning of bracing, could only be tackled in an empirical manner.
- Very little remained within the cables of the specified lubricant and corrosion protection. In consequence individual wires exhibited surface corrosion and pitting and the centres of cables were found to be packed with rust.

The charming Edwardian gondola of Newport bridge is shown in mid-flight in Fig. 11.27.

The transporter bridge at Warrington (built in 1915) has been enhanced

Fig. 11.27 Transporter bridge, Newport, Gwent. The charming Edwardian gondola in mid-flight

to have an improved performance. It was originally designed to carry rail traffic up to 18 tonne, but was modified in 1940 to carry road traffic as well. It was modified again in 1950 to raise the load-carrying capacity to 30 tonne.

RETRACTABLE BRIDGES

Retractable bridges are rather rare and usually small structures of limited span for pedestrians. It is necessary to have large counterweights to balance the movable part during operation. The movement is linear with wheels running on guide rails.

Conservation

The retractable footbridge across the entrance to one of the St Katherine Docks in London was designed by Thomas Telford and constructed in 1829, see Fig. 11.28. The bridge is in two halves, each half having heavy counter-balance boxes containing iron ingots. It was retired from service after 165 years when it was considered inadequate to carry more intensive pedestrian loading attracted by redevelopment of the docks. The two halves were cleaned, painted and repositioned nearby with an explanatory notice. This is an example of preservation rather than conservation, as the bridge is on exhibition but no longer in use.

In contrast to the Telford Footbridge, Shoreham Harbour Footbridge has been retained in service, see Fig. 11.29. The movable section is single-sided across the navigation channel and is part of the thirteen-span concrete structure. It is composed of a steel truss which runs on guide rails and the movement is driven by an electric motor. The opening is approximately 14

Fig. 11.28 Telford Footbridge, St Katherine Docks, London, 1829. A two-sided retractable bridge removed from service in 1994 and preserved *in situ*

Fig. 11.29 Shoreham Harbour Footbridge, 1921. A one-sided retractable footbridge still in service

m. The machinery has been maintained and remains in essentially original condition. Conservation work has been carried out on the concrete and much of the cover has been replaced with sprayed concrete. Nowadays, the movable span is only infrequently operated.

In comparison with other types of movable bridge, retractable bridges are often basic, have fewer moving parts and simpler mechanisms, have less to go wrong, and can require less maintenance.

MANAGEMENT PRACTICE

The actions and strategies available for managing movable bridges can be categorised as:

- Conservation including
    Inspection
    Maintenance
    Enhancement
- Rehabilitation
    Reconstruction
    Conversion (to fixed bridge)
- Replacement

Movable bridges by their nature have greater maintenance requirements than fixed structures, and a conservation-led maintenance strategy requires a broader range of skills and knowledge.

Inspection

Inspection of movable bridges involves work similar to that required for fixed bridges plus the additional requirements of the machinery and moving parts. The intervals between inspections must be shorter, depending on the type of bridge and machinery. It is also necessary to carry out unplanned inspections, as movable bridges are vulnerable to impacts from errant shipping and in some locations suffer frequent collisions and consequential damage.

Monitoring sensitive components of the structure, by measuring torque or local strains, can be used as a supplement to normal inspection to provide early warning of the requirements for repair or maintenance. Such monitoring is via sensors such as strain gauges and inclinometers fitted in appropriate locations. Outputs from the sensors must be processed into a user friendly format and the resulting data transmitted to the bridge engineer.

Inspection requirements are summarised in Table 11.1.

Table 11.1. Inspection activities

Component	Inspection activity
Balance of moving parts	Balance should be checked at appropriate intervals. Balance can be affected by moisture changes in timber components, weight of paint and other maintenance activities involving changing, removal or addition of components.
Decks	Wooden decks must be inspected regularly for signs of damage to the running surface and deterioration of the wooden members. Orthotropic steel decks (rarely original, but sometimes fitted as lightweight replacements) must be inspected regularly for development of fatigue cracks in welded connections to the underside of the deck plate.
Towers and other static superstructure	Wooden superstructures must be inspected regularly for signs of decay, splitting and mechanical damage. Iron and steel superstructures must be inspected for signs of corrosion, broken rivets and bolts, distortion and cracking.
Cables	Moving cables must be inspected regularly for evidence of mechanical wear, fretting, corrosion and fractures of individual wires. Inspection can be done manually. Electromagnetic equipment can aid detection of cracks and broken wires.
Machinery	Inspect moving parts, such as rollers on swingbridges, and all welded connections, for evidence of wear and cracking. Inspect regularly for accumulation of detritus, etc., that can jam moving parts or increase friction and overload drive motors.

Maintenance

Maintenance, like inspection, requires a wider range of knowledge and skills than for fixed bridges. The main activities are summarised in Table 11.2.

Table 11.2. Maintenance activities

Component	Maintenance activity
Moving parts	Ensuring moving parts remain in balance is an essential item of maintenance for all movable bridges.
Decks	Wooden decks can swell, distort and decay. It is necessary to carry out regular maintenance to manage water to avoid leakage and ponding. Waterproof membranes and surfacing must be maintained in serviceable condition. Fatigue cracks in steel decks must be correctly repaired (it is insufficient to replace the weld like-for-like). It may be preferable to remove the deck and carry out repairs off site.
Towers and other static superstructures	Subject to the same maintenance activities as fixed bridges.
Cables	Fixed cables require the same maintenance as on suspended bridges. Moving cables require greasing at regular intervals to minimise friction and wear.
Machinery	Some of the basic mechanical parts may require periodic repair or replacement. When it is necessary to replace historic machinery with modern counterparts it should be retained on site, if possible, as was the case for Tower Bridge.

Enhancement

Enhancement of fixed bridges is generally concerned with raising the load carrying capacity or widening the structure. Enhancement of movable bridges can be targeted at one or more of three objectives.

- Increased load carrying capacity (strengthening)
- Enhanced operational capacity (reduced cycle times)
- Improved maintenance performance (mechanical and electrical components).

In a good example of enhancement of load carrying capacity, the Duluth lift bridge in Minnesota, USA, was renovated in 1985. As part of the renovation, the motors were relocated from the centre to the ends of the span, thereby increasing load capacity available to traffic and extending the life of the structure.

Strengthening to enhance the load capacity of movable bridges will primarily affect the bridge in its static, open to traffic, configuration. Strengthening techniques will be those appropriate to the form of construction as discussed elsewhere. Provided strengthening can be achieved without adding significantly to the weight of a structure, and measures are taken to ensure the distribution of load on to supports whilst opening remains similar, the driving system for the bridge should not necessarily require improvement. However, the opportunity would normally be taken to modernise it if this had not already been done.

In some circumstances, the traffic-carrying capacity can be impaired by loading imposed by the bridge operating machinery and the space it occupies. Relocation of equipment, or use of alternative lighter weight equipment may free up capacity for increased traffic loadings without having to resort to structural strengthening.

A problem with lift bridges has been to ensure that the span rises evenly on both sides. One solution adopted has been to locate the driving machinery in the centre of the span, so that both sides can be driven off a single winch. Modern control techniques can now be installed that allow two or more drives to be controlled to achieve equal rates of lift, and the machinery to be located at the ends of the span or as part of the towers.

As an equivalent measure to 'strengthening' the structural members of a movable bridge, it may be necessary or desirable to enhance the operational capacity of the bridge. On busy routes subject to frequent opening, the capacity of the route over the bridge will be affected by the cycle time. An unusual example of this is the Duluth Aerial Lift Bridge in Minnesota, USA which was originally a transporter bridge (one crossing every five minutes), but was later converted to a lift bridge allowing continuous traffic when closed to shipping.

To a much greater extent than fixed bridges, movable bridges are subject to mechanical wear and tear. Improvements to the operating machinery are a means of enhancing the reliability and reducing the amount of routine maintenance. Methods of enhancement are summarised in Table 11.3.

Conversion to a fixed bridge

On occasions when the need for a movable bridge ceases to exist, due to the closure of an inland port, for example, it is unlikely to be economic, or indeed practical to continue to maintain the bridge in an operational state. Assuming the structure has some historic value, either as a notable example of an engineering technique, or by virtue of the role of the structure in the history of the area, conversion to a fixed bridge should involve the minimum of alteration to the structure and its operating equipment, when balanced against future maintenance and safety requirements.

The operating equipment and mechanisms on movable bridges are varied and often ingenious. This equipment, particularly parts visible externally, are important to understanding how a movable structure would have operated and should ideally be immobilised, but left in place as a feature of the bridge. On the other hand, the fixed bridge will not be subject to the same supervision as the movable bridge so that its equipment and associated pits and chambers may represent a safety hazard and additional maintenance liability if equipment is simply left in place.

Table 11.3. Methods of enhancement

Enhancement	Method
Strengthen to increase load carrying capacity	A common method is to replace the deck with a lightweight steel structure. This has the advantage of minimising knock-on effects on the counter-balance. It has the disadvantage of loss of originality and for a listed bridge it would be unpopular with heritage authorities. The Telford cast iron retractable footbridge in St Katherine's Dock, London, was replaced by a new structure having higher load-carrying capacity. The original structures was cleaned and laid out on the adjacent land as a museum piece.
Convert from manual operation	Small movable bridges on canals and waterways used for leisure activities may be converted from manual operation to manual hydraulic or motorised electrical operation. This has the advantage of added convenience and safety. The disadvantage is loss of originality.
Improved time of operating cycle	With the emphasis on increased speed of movement of traffic and goods, historic movable bridges are often slow to operate and this can present problems. A common solution is to replace the bridge with a modern fast-operating structure with the accompanying disadvantage of the loss of the historic structure. A compromise solution is to retain as much of the original structure as possible and introduce modern components.
Monitoring	Monitoring can be used to enhance performance through improved control of movement operations, for example to control equal movement of independent lifting mechanisms at each end of a vertical lift bridge. Monitoring of torque or tension can give early warning of problems that could overload driving motors. In situations when historic bridges are being operated at higher stresses, monitoring strain gauges or inclinometers can provide early warning of maintenance needs or developing problems, such as cracking.

Consideration may also need to be given to the support arrangement for the fixed conversion of a bridge. Again it must be preferable to maintain the existing arrangement, provided the consequent maintenance liability does not become unduly onerous. Movable bridges often have different support conditions when opening, as compared to when they are carrying traffic, designed to protect the bearings on which the bridge pivots from traffic impact loads.

Where conservation is less important historically, the structure may be modified to remove any redundant parts so as to reduce the maintenance liability. This might apply to ancillary structures, such as operator's kiosks, but could also be more substantial, such as the removal of ballast boxes on bascule bridges. In such cases conservation is clearly not an overriding consideration. There are numerous movable bridges that have been converted to fixed structures, some examples are given in Table 11.4.

Table 11.4. Examples of conversion to fixed bridges

Bridge and date built	Type	Span (m)	Date fixed	Comments
Keadby, 1866 across River Trent	Scherzer bascule	46	1966	Roadway widened, head room increased. The bascule was fastened down.
Hawarden, 1889 across River Dee at Shotton	Swing	43	1971	Machinery removed. A through-truss girder.
Bridgewater Dock, 1871	Retractable (telescopic)	24	c1974	Plate girder rail bridge at 30° skew. Steam boiler removed for preservation. Timber deck replaced by concrete.
Newport, 1934 Middlesborough	Vertical lift	76	1990	A landmark structure having towers 55 m high.

Reconstruction

Reconstruction involves replacement of much of the original structure and is a last resort which would not be appropriate to bridges having historic merit. For working bridges seen as having little merit, such as many of the canal bridges, it is not unusual for them to have been reconstructed several times since the canal was first opened. For example, Zebon Copse Swing Bridge on the Basingstoke Canal was reconstructed in 1954 following collapse, and in 1993 for upgrading. As the original swingbridge was probably built at the same time as other bridges on the canal, around 1792, it is likely that Zebon Copse Bridge would have been reconstructed at least once before 1954. The original brick abutments have been retained so that the latest version is compatible with the environment, see Fig. 11.30.

Reconstruction may be considered to be necessary because the structure has deteriorated to the extent that it is not feasible or economic to carry out repairs, or it may be required to have an enhanced performance which cannot be obtained otherwise.

Reconstruction need not necessarily be a total disaster in a conservation

Fig. 11.30 Zebon Copse Swing Bridge, Basingstoke Canal. Reconstructed in 1954, 1993 and probably in earlier times

sense as it may be possible to design the new structure so that much of the material can be reused. Alternatively it may be practical to preserve major components and move them to a suitable site for permanent exhibition.

The reconstruction can be designed to have the same type of movement, and a structural form that is sympathetic to the original bridge. While this cannot be regarded as conservation in a physical sense, it goes some way towards preserving a link with the original structure. Methods of sympathetic reconstruction are summarised in Table 11.5.

Table 11.5. Methods of sympathetic reconstruction

Method	Comment
Reuse substructure	A practical method well suited to footbridges and canal bridges. Can give a good appearance.
Reuse materials	An environmentally friendly method but not necessarily apparent to the public.
Remove and preserve major components	Only practical for small bridges and footbridges. Carried out successfully on Telford's footbridge, St Katherine Dock, London and Tipton Vertical Lift Bridge (re-erected in the Black Country Museum, Dudley).
Design a sympathetic replacement bridge	Unlikely to be convincing unless some of the original material is reused.

Table 11.6.   Examples of movable bridges

Bridge	Type	Date	Engineer	Navigation width, m	Comments
Selby	Swing	c1791	Jessop	9.1	Reconstructed 1921 and 1970. Has suffered numerous ship impacts.
St Katharine Dock, London	Retractable	1829	Telford		Two-sided. No longer in place, but preserved *in situ*.
Newcastle	Swing	1876	Armstrong	2 × 31.4	Movable section weighs 1300 tonne. Steam pumps replaced by electrical pumps in 1959.
Goole	Swing	1869	Harrison	2 × 30.5	Movable section weighs 650 tonnes. Damaged by ship collisions in 1984 and 1988.
Tower Bridge, London	Bascule	1894	Jones and Barry	61	Closed and refurbished in 1993. Carries high volumes of traffic. Weight restricted.
Duke Street, Birkenhead	Bascule	c1930	Sir William Arrol	34	Refurbished 1993.
Kingsferry	Vertical lift	1960	Anderson and Brown	27	Lifts at 0.02 to 0.3 m/s.
Middlesbrough	Vertical lift	1934	Hamilton and Graves	76	Fixed in 1990.
Newport, Gwent	Transporter	1906	Arnodin	196	Closed in 1985 for refurbishment, reopened 1995.
Middlesbrough	Transporter	1911	Cleveland Bridge Co	172	Crosses the navigation span in 2.5 min. Refurbished in the 1990s.
Shoreham Harbour Footbridge	Retractable	1921	—	9	In essentially original condition.

BIBLIOGRAPHY

Prine, D. (1996) *Remote Monitoring of the Michigan Street Lift Bridge, Sturgeon Bay, Wisconsin, North Western University*, ITI Report 19.

O'Dowd, T. (1995) *Investigation and Reconstruction of Duke Street Bascule Bridge,* Construction Repair March/April.

Jenkins, T. (2000) *Refurbishment of Town Bridge, Northwich, Cheshire,* Bridge Management, 4, Thomas Telford.

Dickson, H. (1994) *Bridges of the Manchester Ship Canal–Past, Present and Future*, The Structural Engineer, **72**(21).

Lark, R.J., Mawson, B.R. and Smith, A.K. (1999) *The Refurbishment of Newport Transporter Bridge*, The Structural Engineer, 77(16).

Groome, L.W., Halse, W.I., Longton, E.H. and Stephens, D.L. (1985) *Tower Bridge*, The Structural Engineer, 63A (2).

Chapter 12

Concrete

ABBREVIATIONS

SSSI	Site of Special Scientific Interests
ASR	Alkali Silica Reaction
PFA	Pulverised Fuel Ash

GLOSSARY

Mass concrete	Concrete having no reinforcement.
Reinforced concrete	Concrete reinforced with steel bars. This is distinct from iron frames clad in concrete.
Prestressed concrete	A generic term used loosely to refer to several systems where the concrete is held in compression by stressed steel.
Precast prestressed concrete	Concrete cast in moulds having pretensioned steel, usually prepared in factory conditions.
Internal post-tensioned concrete	Precast concrete post-tensioned by steel pushed through internal ducts, tensioned and locked off. The ducts are usually filled with grout under pressure.
External post-tensioned concrete	Precast concrete post-tensioned by steel external to the concrete section.
Strand	A group of wires spun in a helical form around a longitudinal axis formed by a King wire. Common types are 7-wire and 19-wire.

Cable	An assemblage of external strands or bars.
Macalloy bar	High strength prestressing bar (trade name).
Tendon	Generic term for post-tensioning steel, bar or strand.
Segmental construction	Longitudinal beams cast in separate lengths (segments) and post-tensioned to form a continuous element, can be precast or cast *in situ*.
Alkali silica reaction	A chemical reaction in hardened concrete which can cause expansion, cracking and exudation of a gel.
Pulverised fuel ash	A pozzolanic material used as a partial replacement for Portland cement.

BACKGROUND

Concrete bridges were first built in the late 1800s and there are now many early examples that are up to 100 years old and in continued use. Some have been heritage listed and others have historic significance and should be treated as having similar status.

As a material, concrete has been steadily developed to have improved properties such as higher strength, faster curing times, lower permeability, etc. In consequence, attention tends to be focussed on latest developments and concrete continues to be regarded as a relatively modern material. Less attention is paid to early concrete and its characteristics are less well known.

Initially concrete was introduced as cladding to improve the fire resistance of iron and steel frames. It was also used to provide corrosion protection. The structural attributes of concrete were quickly recognised and bridges were designed in mass concrete, reinforced concrete and in later years, prestressed concrete. The structural forms of *in situ* concrete were developed from mass concrete arches to open spandrel arches, beam-and-slabs, etc. The common forms of structure are summarised in Table 12.1.

When carrying out conservation work, it is necessary to be aware of the different systems of reinforcement used in the early years of concrete, for example Hennébique, Coignet and Considère, of which Hennébique was most commonly used in Britain. Also there were various types of reinforcement bar having different profiles and strengths supplied by companies such as the Trussed Concrete Steel Company, Indented Bar and Concrete Engineering Company, the Perfector system, and others (detailed in *The Engineers' Year Book* of 1923). The properties and characteristics of these systems are factors that have to be taken into account when assessing load carrying capacities to current standards.

Prestressed concrete was introduced in Europe in the 1930s and a little later in Britain. Bridges having precast prestressed beams with *in situ* decks quickly became popular as being an economic and convenient form of construction. Post-tensioned concrete was developed using various stressing

Table 12.1. Structural forms of concrete bridges

Structural form	Description	Appearance (elevation or cross-section)
Arch	Monolithic structure composed of mass concrete or structure having closed spandrels with fill beneath deck as in traditional masonry arch, and slab deck or structure having closed spandrels and internal walls (or vaults) supporting slab deck.	
Open spandrel arch	Deck supported on piers connected to arch barrel or separate ribs.	
Beam-and-slab	Superstructure having longitudinal beams and cross-beams, usually cast monolithically with deck slab. The longitudinal beams sometimes have curved soffits to resemble flat arches.	
Slab	Slab deck reinforced with conventional reinforcement.	
Bowstring	Deck supported on transverse beams suspended by hangers connected to arch ribs (known as rainbow arches in the United States).	

Table 12.1.   Continued

Structural form	Description	Appearance (elevation or cross-section)
Portal	Monolithic structure having supports and deck cast monolithically *in situ*.	
Filler beam slab	Slab deck reinforced with I-beams, often having no transverse reinforcement.	
Jack arches	Composite structure with brick arches.	
Jack arches	Composite structure with buckle plates.	
Trough deck	Composite structure with steel troughing.	
Precast concrete	Half-through, panel construction.	
Precast beams	Precast prestressed standard beams.	WR SBB Inverted T

systems, for example Freyssinet, Magnel Blaton, Gifford-Udall, etc. (see Andrew and Turner, 1985). An early example of post-tensioning is Waterloo Bridge, London, see Fig. 12.1. Construction began in 1937 and was completed in 1944. The structure is basically reinforced concrete having prestress applied in certain areas to enhance its shear capacity. The bars were

Fig. 12.1 Waterloo Bridge, London, 1937–1944

stressed by steam heating and were then constrained with turnbuckles to obtain a modest prestress of 200 N/mm^2.

Some examples of concrete bridges in Britain are given in Table 12.6 at the end of this chapter.

The impact of utilities

Utilities have generally had less impact on concrete bridges than on masonry arches, as they have invariably been accommodated at the time of construction. On occasions when utilities have been added afterwards the appearance of the bridge has often been spoiled. The attachment of external utility pipes presents problems due to the added weight and the need to fix bolted connections to the concrete. Drilling bolt holes into reinforced or prestressed concrete requires great care to avoid damage to the steel and consequential weakening of the structure.

In an extreme case, a 300 mm diameter steel gas pipe was attached to the side of a grade II listed post-tensioned concrete footbridge. The pipe was attached by bolted brackets fitted at intervals along the web of a main beam completely spoiling the appearance of the side elevation. During installation the bolt holes were drilled through the post-tensioning ducts in five places severing the steel tendons. The resulting loss of prestress weakened the structure and generated longitudinal cracks, which extended and widened progressively. This damage was not discovered until some years later. Due to the geometry of the bridge it was not possible to repair or strengthen the beam. The gas pipe was removed and repositioned beneath the bridge, but the bridge itself had to be demolished and a new one was built.

Performance of concrete

The early bridges were constructed with materials and detailing now regarded as having poor durability. Moreover, loading requirements were much lower and many of the bridges have had to be strengthened to meet current standards. The concrete had higher water/cement ratios and thicknesses of cover were low, see Fig. 12.2. In consequence of these shortfalls, carbonation depths have developed over 50 or more years of service to the point where they now exceed the cover thickness in many cases so that reinforcement is no longer protected by alkaline concrete. The concrete invariably contained voids and honeycombing due to the poor methods of compaction employed before the introduction of vibration. Despite these shortfalls, early bridges have performed surprisingly well and comparatively few have had to be demolished due to gross deterioration. In contrast, bridges of the motorway era seem to have performed less well and some have developed serious corrosion in as little as 20 to 30 years, albeit it is difficult to make objective comparisons because of the different circumstances. In one of the few comparative research studies to be carried out, it was found that concrete in bridges built before 1960 exhibited properties regarded as being related to durability that were significantly poorer than those of more recent concrete bridges (Brown, 1987). However, chloride levels were found to be lower and the condition of the concrete was generally good, in contrast to expectation. The findings were based on measurements of cement content, water/cement ratios, oxygen permeability, capillary porosity, water absorption and indirect tensile strength.

The most common maintenance problem with concrete highway bridges has resulted from corrosion of the reinforcing steel, due to the application of de-icing salt during cold spells. Some of the early bridges were constructed using marine aggregate that was either inadequately washed or not washed so that there were chlorides present from the start. The performance of reinforced concrete bridges has been variable as some have suffered comparatively little corrosion whereas others have been badly affected. Dor-

Fig. 12.2 Low thickness of cover and consequential corrosion

nie Bridge, Scotland, is an interesting example as it developed significant cracking, spalling and corroded reinforcement. As the assessed loading capacity was only 17 tonnes, it was decided that it would have to be replaced. The bridge, built in 1940, was a 16-span beam-and-slab structure having a total length of 230 m. Before demolition, sections of the deteriorated bridge were load tested to collapse by TRL. The load tests were carried out on a cross-beam, deck slab and two main beams. Calculated failure loads, prior to the tests, predicted that the cross-beam would fail at 760 kN (failed at 2,530 kN), the deck would fail at 470 kN (failed at 2,910 kN) and the main beams would fail at 1,130 kN (loaded respectively to 5,180 kN and 5,310 kN, but not failed).

In the past, many reinforced bridges have been unnecessarily replaced due to shortfalls in their assessed strength but, unlike Dornie, exhibiting no signs of distress.

In contrast to the problems of reinforced concrete, mass concrete bridges such as Axmouth Bridge in Devon have benefited from having no reinforcement to corrode, see Fig. 12.3.

Precast, prestressed beams are a popular form of construction available in standard sections since the late 1940s. These include the WR beam introduced in 1948, the SBB beam at about the same time and still in production, and the inverted T-beam in 1951, also still in production. Outlines of the cross-sections of these beams are given in Table 12.1.

Precast, prestressed beams have performed well over a period of up to 50 years and there have been no serious problems. When in 1989, a survey of the condition of 200 concrete bridges was carried out for the Department of Transport (Wallbank, 1989), no problems or signs of distress were reported for the prestressed beams in seventy-four bridges of that type. The excellent durability of prestressed beams is almost certainly due to a combination of factors: high quality control exerted in the factory con-

Fig. 12.3 Axmouth Bridge, Devon, 1877. Mass concrete construction

ditions of precasting yards, good quality concrete having consistent cover to the prestressing steel, and good detailing of the standard beam designs.

In the early years, shear reinforcement was not required by the design codes of that time, albeit stirrups were often used to hold together the reinforcement and ducts during construction. There were, however, beams having no effective shear reinforcement which were deemed to be insufficient when assessed in later years, and were demolished.

The performance of post-tensioned concrete has been less satisfactory than pretensioned precast concrete. Grouting has been the Achilles heel of post-tensioning and there have been numerous cases where apparently well-grouted bridges have been found to have partially grouted or empty ducts. This has been a common experience in most countries where ducts have been inspected and is not confined to the early bridges. The absence of grout does not necessarily lead to problems and there have been many occasions when the steel has been found to be exposed, but in good condition. There have also been occasions when the steel has corroded and fractured. In an extreme case Ynys-y-Gwas Bridge (built in 1953) collapsed in 1985. Although it has been suggested that the bridge was unique, the collapse was due to several factors that have sometimes been found in other post-tensioned bridges, namely, that some of the ducts had not been grouted properly and had large voids, chlorides from de-icing salt had leaked through a waterproofing layer and the joints between precast beams and beam segments, and severe local corrosion of the post-tensioning tendons occurred without exhibiting any external evidence. There have also been collapses of post-tensioned bridges in Belgium, Switzerland and Japan, and in Britain a number of post-tensioned superstructures have been replaced due to having corroded tendons.

CONSERVATION OF MASS AND REINFORCED CONCRETE

Concrete bridges are commonly repaired and strengthened, but rarely treated to conservation. Whereas conservation of the traditional materials timber, stone, iron and steel, has attracted considerable research and development over the years, concrete has not generally been regarded as a heritage material. There has, however, been considerable research into maintenance of modern concrete much of which is relevant and applicable to old concrete. Attitudes to conservation of concrete are changing as more concrete structures become listed as national or international monuments. There have been examples of conservation of concrete bridges in recent years involving techniques that do justice to historic concrete as described in the following sections and summarised in Table 12.2.

Grout voided concrete

At the time when older concrete bridges were constructed, vibrators were not available so that it was necessary to have higher water contents to aid workability and use tampers to compact the concrete. This could result in honeycombing and voids. When there is a degree of interconnection between the voids the concrete can be made good by injection of a flowable

Table 12.2. Methods of conservation of mass and reinforced concrete

Method of conservation	Advantage	Disadvantage
Grout voided concrete	Strengthens the defective concrete without change to its external appearance.	Unsuitable when voids are disconnected. Care needed to avoid unsightly leakages and spalling.
Seal cracks	Restores protection to the concrete and steel without change to its external appearance.	Unsuitable for live cracks. Necessary to ensure that cause of cracking is fully understood.
Penetrant coatings	Colourless coatings can consolidate deteriorating concrete and provide hydrophobic protection. No change to appearance.	Not suitable for some types of concrete.
Paint	Provides a uniform surface appearance when repairs or weathering has made the concrete unsightly.	A last resort which changes the external appearance. Attracts graffiti. Can crack, spall and become unsightly with age. Requires future maintenance.
Corrosion inhibitors	Easy to use, can be applied to concrete surface. It is claimed that the rate of corrosion can be reduced to 10%.	Accurate control of the dosage is important. Preferable to monitor performance. May require regular application. Ineffective if concrete surface is impermeable.
Replace cover (patch repair)	Restores cover and protection to the steel reinforcement	Likely to change the appearance unless special care is taken. Poor preparation or use of incompatible materials could worsen the situation if chlorides are present.

cementitious grout. This was carried out successfully on Axmouth Bridge, Devon, using a grout formulation of Portland cement and PFA in equal proportions, plus an expansive admixture. Care was taken to control the pumping pressure to avoid blowing off the vulnerable surface layers of the con-

Table 12.2. Continued

Method of conservation	Advantage	Disadvantage
Sprayed repairs	Commonly used method to place hard dense mortar	The as-sprayed finish can have a rather poor appearance that does not match *in situ* concrete in either texture or colour. When reinforcement bars are present, there can be a shadow effect where small voids are created behind the bars. Requires skilful workmanship.
Replace defective components	Can be carried out like-for-like, so that there is no change in the appearance. Replacements can be made stronger and more durable than originals.	The work can cause disruption to traffic unless temporary measures are taken such as propping. Heritage authorities are sometimes reluctant to agree replacement of original material. Expensive.
Expose and paint steel framework	Reduces dead load and raises load carrying capacity. Economic.	Applicable to a limited number of structures. Reduces originality and changes the appearance.
Pedestrianise	An alternative to strengthening which enables the bridge to be preserved in its original form.	Can attract a lower priority for subsequent maintenance work.

crete. This enabled the attractively weathered appearance of the bridge to be retained, see Fig. 12.3.

Grouting is not always successful and is not suitable for concrete having voids that have insufficient interconnection.

Crack sealing

On occasions when concrete has become cracked, it may be appropriate to seal the cracks with a low viscosity resinous sealant. The main purpose is to prevent the entry of water that could lead to frost damage to the surface concrete and corrosion of the steel reinforcement. The sealant should be colourless, so that there is no change to the appearance of the concrete.

Crack sealing is only appropriate when the cracks are stationary, i.e. their widths are not increasing progressively with time due to some form of structural deterioration.

In practice, cracks are unlikely to be totally stationary as temperature effects will cause cyclic behaviour. Flexible sealants are claimed to be able to cope with such movements, provided they are not progressive. Crack sealing is appropriate when corrective actions have been taken to prevent further movement or when underlying movements, such as uneven settlement, have stabilised. Examples of bridges having crack sealing include the Royal Tweed Bridge at Berwick.

Penetrant coatings

The actions of penetrant coatings are briefly described in Chapter 7 in relation to masonry, where they can be used to halt the processes of deterioration and erosion. Concrete presents a somewhat different proposition as it is generally stronger and less permeable than stonework. The coatings can be designed to strengthen (consolidate) weak material or act as water repellents (hydrophobes), the latter are relevant to concrete. There are many types of penetrant coatings and their effectiveness is yet to be fully established.

The requirement of a concrete coating is to be hydrophobic and prevent ingress of water and soluble contaminants such as chlorides. Hydrophobic coatings permit moisture to escape so that the concrete can dry out and cause corrosion activity to slow down or stop altogether. Such coatings may therefore be effective in cases when corrosion of the reinforcing steel and deterioration of the concrete has not advanced to the stage when it is necessary to replace the cover concrete. Hydrophobic coatings, such as silanes, are applied to new concrete structures, repaired structures and existing structures, where there are no significant levels of chloride contamination.

Penetrant coatings need not affect appearance of the concrete as they are usually colourless. However, experience on stone indicates that they deteriorate after about 15 years and retreatment is required. Application to concrete is fairly recent and a method of determining when to re-treat has not yet been faced, let alone standardised. Likewise, the problems of re-treatment have not been assessed.

Corrosion inhibitors

Corrosion inhibitors have been widely used to protect steelwork and, less commonly, to protect steel in concrete. They can be applied to surfaces of existing concrete or used as admixtures in the concrete mix.

When applied to an existing structure, the inhibitor must diffuse through the concrete to the level of the reinforcement in order to be effective. With liquid inhibitors this can take some time as diffusion of vapour phase inhibitors is more rapid through dry concrete. There are three main types of inhibitors: anodic inhibitors, cathodic inhibitors and multifunctional inhibitors.

Anodic inhibitors work by retarding the reaction at the anode, calcium nitrite is the most widely used anodic inhibitor, though other nitrites have been used. Dosage rates are very important as all anodes must be suppressed, otherwise localised pitting attack may occur at those remaining. Calcium nitrite is consumed with time and only works above a given activation level (i.e. chloride/nitrite ratio) at which point the inhibitor must be reapplied or corrosion will recommence.

Cathodic inhibitors retard the reaction of the cathode. Benzoates, particularly sodium benzoate, are the most effective cathodic inhibitors and have the advantage that they are not consumed with time. However, inhibition takes longer to establish and the compressive strength of the concrete is reduced.

Both anodic and cathodic inhibitors therefore have disadvantages and multifunctional inhibitors have been used to combine the benefits of both anodic and cathodic inhibitors. Multifunctional inhibitors are based on amino alcohol.

It should be noted that researchers are divided on the effectiveness of corrosion inhibitors and it has even been suggested that they can worsen the situation, although this is an extreme view. However, when applied as admixtures in the concrete mix, they appear to be reasonably effective.

Paint

Paint is sometimes applied as a cosmetic to concrete that has defects or unsightly repairs. An example of a well-painted bridge that has remained in good condition after several years is given in Fig. 12.4. Paint can impart a bright and shining finish that is not compatible with old concrete and will eventually degrade under the UV content of sunlight. Its ability to deform

Fig. 12.4 Woodbridge Old, Guildford, 1912. After being strengthened and refurbished, the concrete was treated with an elastomeric paint which was in pristine condition after some nine years

and bridge cracks is limited to tolerable widths so that cracks can reflect through the coating and produce an unsightly effect. Unsightly biological growth can develop on shaded areas. Vandalism and graffiti are likely to occur on all types of structure, but tend to look worse on clean and bright, painted surfaces. Biological growth and graffiti are relatively easy to clean off, although this becomes a burden for maintenance.

Current policy of the Highways Agency is not to paint concrete unless there are overriding engineering or environmental reasons.

Unfortunately, painted concrete requires maintenance and towards the end of its life is likely to become distinctly scruffy. The paint embrittles with age, cracking develops and the paint peels back as shown in Fig. 12.5. The going life of paint is about 15 years and longer for some systems.

Replace defective cover (patch repairs)

It is not uncommon for the cover concrete to require replacing. This is due to the fact that over the years the concrete can deteriorate due to actions such as freeze–thaw cycling, ingress of chlorides and carbonation. Corrosion of the reinforcement can cause cracking and spalling. However, simply replacing the cover concrete could be a waste of time, because it is necessary to deal with underlying problems such as deeper penetration of chlorides and corrosion of the reinforcement. Concrete containing significant levels of chloride must be removed and the reinforcing steel thoroughly cleaned so that all corrosion products are removed. Most importantly chlor-

Fig. 12.5 The appearance of paint towards the end of its life. Despite the best efforts of a graffiti artist, this abutment has become unsightly and overdue for repainting

ides must be removed from the corrosion pits. If this is not done properly, corrosion is likely to recur and the repair material will be forced off in a short time. There are occasions when it may be appropriate to remove chlorides by electrochemical methods, as mentioned later in this chapter.

Repair materials are commonly flowable mortars that are trowelled or cast in place. These materials often provide a poor colour match giving an ugly appearance as exemplified by Fig. 12.6. Moreover, if the underlying cause of the problem is not correctly identified, the repair concrete may crack as before.

As an example of patch repairs, an open spandrel arch bridge across the River Fnjöska in Iceland, constructed at the beginning of the century, has special historic significance as being the first reinforced concrete arch in the country. By the early 1990s, its surface concrete had deteriorated and reinforcement bars were corroding. It was therefore necessary to replace defective concrete with material that matched the weathered appearance of the bridge. A mortar was formulated containing Portland cement, polymer fibres and microsilica. A polymer-modified additive mixed with local sand produced the required colour, texture and physical properties. The defective concrete was removed and the reinforcement cleaned to expose bright steel. A protective coating was applied to the reinforcement and the new mortar placed in thicknesses up to 75 mm.

King George VI Bridge, Aberdeen, Scotland, is a three-span granite-faced reinforced concrete arch bridge across the River Dee, constructed in 1941. By the late 1980s, carbonation had developed in the concrete soffits, the reinforcement was corroding and concrete was cracking and spalling. Patch repairs were carried out by removing defective concrete, cleaning the reinforcement and applying an acrylic mortar. To minimise the risk of further deterioration, an anti-carbonation coating was applied to the soffits.

Fig. 12.6 Patch repairs. Repairs made using an ill-matched mortar and without dealing with the root cause of the original cracking

Patch repairs to the landmark Dumbarton Bridge across Rock Creek Parkway in Washington, USA are shown in Figs 12.7 and 12.8 in progress and afterwards.

Sprayed repairs

Sprayed repairs are a variation on patch repairs. Whereas with a patch repair the material is trowelled or cast in place, sprayed concrete is forcibly projected into place, see Fig. 12.9. Spraying is rather less precise than patching and is typically used only on high volume repairs where the depth to be filled, and particularly the area to be covered, are large.

There are two common materials used in sprayed repairs: concretes and mortars (i.e. a concrete with no coarse aggregate). These are 'shotcrete' and 'gunite', respectively. The terms are used rather loosely and to avoid misunderstandings (here and in Chapter 7) mortar spraying is referred to in relation to conservation and concrete spraying for strengthening. The material can be applied using wet process spraying or dry process spraying. In wet process spraying, a pumpable mix is forced under pressure through

Fig. 12.7 Dumbarton Bridge, Washington. Patch repairs being carried out

Fig. 12.8 Dumbarton Bridge, Washington. Bridge after repair

a nozzle on the end of a hose and the rate of placing is dependent on the rate of delivery from the batching plant.

In dry process spraying, the repair material is dry-batched and mixed in a conventional plant. Water is added at the nozzle by the operator who can adjust the rate at which it is added and so control the consistency of the mix to maximise adhesion, and minimise rebound or slumping after spraying. Typical strengths that can be achieved with dry process spraying are comparable with those of a good quality *in situ* concrete, but because the water is added on site, the final mix is prone to be more variable than with wet process spraying which uses a uniformly consistent mix.

Whichever process or material is used the existing substrate should be dry and precautions observed regarding preparation of any exposed reinforcement as for patch repairs. If there is little or no original reinforcement exposed, it will be necessary to reinforce the repair with secondary reinforcement, either in the form of lightweight mesh or reinforcement bars fixed to the substrate. This strengthens the repair, and the mechanical interlock helps with adhesion and reduces slump. However, care must be taken when spraying to avoid the so-called 'shadow effect' where voids are left

Fig. 12.9 Repairing by sprayed concrete

behind the reinforcement as viewed from the nozzle. In both the wet process and the dry process the skill of the operator is crucial to a good result.

Sprayed repairs to concrete have been used on many occasions, for example Shoreham Harbour Footbridge was gunited after about 50 years in service and has remained in good condition for a further 30 years, see Fig. 12.10.

Salginatobel Bridge, Switzerland, designed by Maillart and constructed in 1930, is listed as an International Historic Civil Engineering Landmark. When defective cover had to be replaced, conservation issues and retention of the character of the concrete were foremost in the requirements. Surface concrete was removed by high pressure water jets to a depth of 10 to 20 mm from most of the visible areas and replaced by at least 30 mm of shotcrete to ensure adequate thickness of cover. The replacement material was applied by spraying and applying shuttering to the fresh material to provide a surface finish that was compatible with the rest of the bridge.

Replace defective components

Structural (load-bearing) components that have become defective for one reason or another can be replaced on a like-for-like basis without changing the appearance of the bridge beyond the introduction of a freshly cast and new section of concrete. This would normally be acceptable because the boundaries between new and old concrete can be formed by the structural connections as opposed to the ragged shapes of most patch repairs and if not, straight geometric boundaries can be contrived. Structural, load-bear-

Fig. 12.10 Shoreham Harbour Footbridge, West Sussex, 1921. The concrete superstructure was successfully repaired by gunite

ing components that most commonly have to be replaced include decks, piers and cross-heads.

Replacement of components provides an opportunity to raise strengths to meet current standards and provide greater durability. In the latter context, more durable concrete can be placed having a lower permeability, higher thickness of cover and better compaction. Matching of the appearance of the concrete, colour and surface texture, are rarely considered as an important issue, while factors such as strength are usually dominant. For components such as decks, this may not be too important as the side elevation is dominated by parapets and other edge details which may remain unchanged. For other components, substitution of modern concrete could spoil the appearance of the bridge.

Examples of bridges having replaced components include Pelham Bridge, Lincolnshire, a multispan structure having a total length of 350 m and constructed in 1957. Leakage through the expansion joints led to high chloride levels and damage to the concrete piers. Twelve of the piers were replaced with precast elements up to 4.6 m long and 8 tonne in weight, c1980. A further sixteen piers were refurbished by removal of defective concrete cover, cleaning the reinforcement bars and refacing with super-plasticised concrete pumped into formwork under pressure.

Hillhurst (Louise) Bridge across Bow River in Calgary, Canada is a five-span reinforced concrete arch structure constructed in 1921. It has four lanes carrying vehicles up to legal limits of 63 tonne. Concrete in the spandrel walls and arch edges suffered severe deterioration due to poor drainage, the action of de-icing salts, freeze–thaw damage, and to a lesser extent, alkali–silica reaction. A refurbishment scheme was designed to meet requirements to retain the original historic architectural appearance. Among other things, the deck was replaced by precast prestressed panels over-topped with *in situ* concrete, and precast cantilever brackets were installed to support the footways on either side of the bridge. The work was completed in 1995.

Baltimore Street Bridge across Gwynns Falls in Baltimore, USA is a three-span open spandrel arch structure constructed in 1932. Inspections revealed that the deck was below the strength required for modern traffic and the concrete was badly deteriorated having high chloride levels, cracking and exposed reinforcement. The bridge was treated as a historic structure so the repair work had to meet the approval of the State Historic Preservation Office (SHPO). The structure above the extrados of the arch ribs was replaced to the requirements of the SHPO, namely the elevation viewed from below remained unchanged, unsightly patch repairs were to be avoided where possible, and the *in situ* concrete was to have a uniform appearance and texture. The work was completed in 1999.

Expose and paint steel frames

One of the early forms of construction comprised steel frames encased in concrete. The purpose of the concrete cover was to provide fire resistance and protect the steel frame from corrosion (this was at a time when concrete was believed to provide infinite protection against corrosion).

The high level Farnham Road bridge carrying the A31 across the A3 at Guildford, Surrey exemplifies this form of construction. The bridge, constructed in 1933, is an open arch having a span of 25.5 m. The arch and piers are riveted steel beams originally encased in concrete reinforced by lightweight steel mesh.

By the 1990s, the concrete had become cracked and rust-stained. Some pieces fell off causing a hazard to motorists passing below, but fortunately there were no accidents. Structural analysis showed that removal of the concrete cover had no adverse effect on the strength, in fact the reduction in dead weight raised the load-carrying capacity from 38 to 40 tonne. The exposed beams were cleaned and painted.

On a historically more significant bridge this treatment might not be acceptable to heritage authorities because of the loss of originality and changed appearance. On the other hand, it could be argued that exposure of the steel frame produced a structurally more honest solution that is an improvement on an original design based on a fallacious assumption. Certainly the new appearance is light and attractive as shown in Fig. 12.11.

Pedestrianise

When concrete bridges are found to have inadequate load-carrying capacity they are usually strengthened by one of the methods described in a later section of this chapter. However, bridges having historical significance, where heritage authorities would be reluctant to permit such changes, are sometimes downgraded from vehicular traffic to pedestrians. Examples include Axmouth Bridge in Devon, Fig. 12.3, which was by-passed by a new bridge and Homersfield Bridge, in Suffolk, Fig. 4.12, which became part of a bridleway following construction of a by-pass in 1971.

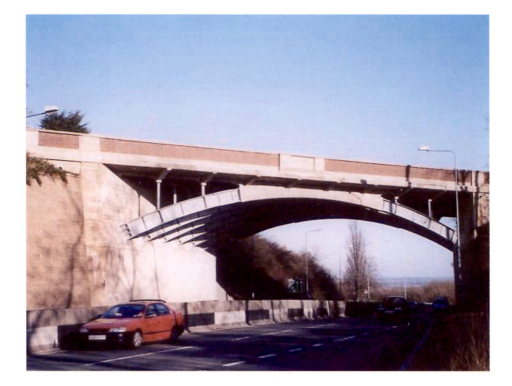

Fig. 12.11 Farnham Road bridge, Guildford. Cracked and deteriorated concrete was removed and the steel frame exposed and painted

CONSERVATION OF PRESTRESSED CONCRETE

There are only a small number of prestressed bridges that have reached an age and stature where they are recognised as meriting conservation work. To date, the prestressed components have required little maintenance with the exception of post-tensioning systems. Some post-tensioned bridges have deteriorated and had to be demolished as, for example, the historic Wallnut Lane Bridge in the USA. Where it has been feasible to carry out repairs they have generally involved designs that have had little impact on the external appearance and would meet the requirements of conservation if they had been imposed. This has been a consequence of the geometry and location of the post-tensioning which has favoured the more pleasing schemes. Methods of conservation of prestressed concrete are summarised in Table 12.3.

Regrouting

During inspections, significant voids are occasionally discovered in the ducts. The voids may be relatively small or they may extend for most of the length of the duct. On occasions when the exposed strands have corroded and fractured it is necessary to investigate whether the problem is widespread in which case it is likely to be necessary to replace the deck altogether. On occasions when the exposed strands are only lightly corroded, or free of corrosion altogether, it is necessary to decide on an appropriate strategy, the alternatives being to let well alone, take no immediate

Table 12.3. Methods of conservation of prestressed concrete

Method of conservation	Advantage	Disadvantage
Regrouting voided or empty ducts	No changes to the appearance.	Requires experience and skill to carry out. Several bridges have been regrouted but experience is limited.
Monitor the post-tensioning system to detect fractures	A non-destructive system that can be operated remotely. Used successfully on several structures. No change to the appearance.	Good experience, but limited number of applications to date.
Bonded plating	An economic and effective method. Used successfully on numerous bridges.	The steel can corrode at the interface with the concrete and requires inspection at regular intervals. Changes appearance of the soffit.
Replacement of external post-tensioning tendons.	Can be carried out with minimal interruption to traffic. No change to appearance. Improved systems of protection can be installed.	Depending on the bridge in question, the anchorages may pose difficulties.

action but monitor behaviour, or regrout. The decision must of course take account of the sensitivity of the design to loss of prestress and likelihood of collapse.

Monitoring has been carried out successfully on a number of occasions and is discussed further in the section on 'Management practice'.

Regrouting is a difficult operation which has been carried out on a number of occasions sometimes using vacuum assistance. To be successful it is necessary to identify the locations and sizes of all the voids beforehand. The quantities of grout injected into the voids should be measured and checked against the estimated requirements. To date, there is limited experience of regrouting and its efficacy remains to be established.

Bonded plating

The method of bonded plating is described later in the section on strengthening.

Some of the early bridges having prestressed longitudinal beams were post-tensioned transversely to ensure adequate distribution of the wheel loads. In designs not having *in situ* slabs cast on to beams, surfacing was laid directly on to the top flanges, so that there was a potential route for leakage between the beams. In bridges constructed to this design, the transverse prestressing steel is susceptible to corrosion and fracture. In some of these cases it has been possible, by load testing, to demonstrate that there is adequate transverse load distribution despite the loss of prestress. This was for shorter span bridges where the fixity at the ends of the deck was sufficient to provide the necessary lateral continuity.

There have been other occasions when it has been considered necessary to provide added reinforcement using bonded plating, for example, the Seckenheim-Ilvelsheim Neckar Bridge in Germany has three bowstring spans of 56 m. One of the spans was reconstructed in the late 1940s having a prestressed concrete slab suspended from the arch beams. In the late 1980s, following the discovery of 0.1 mm wide cracks in the soffit, the transverse post-tensioning ducts were invasively inspected and it was found that one third of the tendons were either not grouted or partially grouted. The prestressing steel was severely corroded and exhibited low fatigue strength. A strengthening scheme was designed having bonded transverse steel plates.

Replacement of external post-tensioning

External post-tensioning has the advantages of being inspectable and replaceable. However, durability has been variable even for cables located within box beams where the environment should be free of corrosion.

Two bridges in Britain namely the Braidley Road Bridge, Bournemouth, Dorset and the A3/A31 Bridge, Guildford, Surrey, had external strands, protected by polymer coatings and located within box beams. In both cases there were occurrences of fractured wires and strands and it was necessary to replace the cables. The fractures were detected from the rumpled appearance of the coatings. Replacement of the strands was achieved with minimal interruption to traffic.

Great Naab Bridge, Bavaria, Germany is a continuous two-cell box beam structure having three spans of 27.2 m, 340 m and 27.2 m. There are diaphragms at each midspan and over the piers. The bridge was constructed in 1954. Within 2 years a deflection of about 30 mm developed in the centre span and cracks of up to 0.35 mm width on the soffit. Additional post-tensioning was put in place using sixteen strands of 38 mm diameter to reinstate the initial prestressing force. The cables were protected by an anti-corrosion paint. After 21 years, these additional cables were found to be badly corroded and one had fractured. Corrosion was worst where the cables passed through the diaphragms and it was impossible to renew the corrosion protection. It was necessary to replace the cables using an improved system of protection and, where they passed through the diaphragms, the spaces were filled with grout. These examples of failures and rehabilitation of externally post-tensioned bridges illustrate on the one hand that great care must be taken in providing good corrosion protection. On the other hand, such construction is easier to inspect than internal grouted post-tensioning and cables can be replaced more easily and without requiring full closure to traffic.

ELECTROCHEMICAL METHODS FOR TREATING CORROSION

The electrochemical methods for treating corroding reinforcement are a specific form of conservation, and it is convenient to treat them as a separate issue (Broomfield, 1997). They are summarised in Table 12.4.

Cathodic protection

When reinforcement corrodes, it acts as an anode in an electrochemical cell. Metal ions pass into solution as positively charged hydrated ions and the excess electrons flow through the metal to cathodic sites where electron acceptors can consume them. The principle of cathodic protection is to make the reinforcement cathodic, so that corrosion cannot occur.

There are two types of cathodic protection systems available: those using sacrificial anodes, and those using impressed current systems. Both techniques require only the repair of spalled or delaminated concrete, as adjacent undamaged concrete does not need to be replaced, even if heavily contaminated with chlorides. Sacrificial anodes are less effective for use with reinforced concrete.

For cathodic protection using an impressed current, an anode can be embedded or attached to the surface of the concrete and a constant voltage is applied between this and the reinforcement, which acts at the cathode. The cathodic reaction produces hydroxyl ions that increase the alkalinity of the concrete surrounding the cathode. Negatively charged ions are repelled by the cathode and migrate to the anode at the surface of the con-

Table 12.4.   Electrochemical methods of treating reinforced concrete

Method	Advantages	Disadvantages
Cathodic protection	An effective method of halting corrosion used on bridges in the USA and elsewhere for many years.	Requires regular attention and monitoring. Danger of triggering ASR in susceptible concrete. Application of the anode can change the appearance. Some areas may be difficult to access. Expensive
Re-alkalisation	An effective method of restoring protective properties of the concrete.	As above.
Chloride removal	Treats the cause of deterioration (corrosion) rather than the symptoms.	Risk of triggering ASR in susceptible concrete. Can increase porosity of the concrete. Requires retreatment from time to time.

crete. Impressed current cathodic protection has quite a long history, having been used on concrete bridge decks in the USA for many years.

The anode can take a variety of different forms. Conductive coatings applied to the surface, such as carbon-based paints, have been used but early versions tended to suffer from poor durability and premature failure, due to flaking and peeling, although they are relatively cheap to install and easy to repair. Some coatings have the added advantage that they provide a waterproof barrier as well. Sprayed zinc coatings have been used in marine environments in the USA for some time, though zinc is toxic and its use raises environmental concerns.

A conductive mesh, sometimes coated with a conductive polymer or plated with platinum, and embedded in a layer of cast concrete (on a horizontal surface) or sprayed layer of concrete (on vertical surfaces or soffits) is another common form of anode. Materials used for the mesh include copper-, titanium- and carbon-based materials.

Because concrete is a variable material and chloride levels and rates of corrosion will not necessarily be constant for all parts of a structure, larger structures are usually divided into zones. Each zone requires a separate power supply and associated monitoring and control system to regulate the applied voltage. This increases both the capital and maintenance costs of the system.

Even within a zone, local differences in potential between the anode and cathode mean that the applied voltage, which has to be large enough to stop corrosion at the most vigorous anode, will be larger than required for the other anodes in the zone. This overprotection can lead to the production of hydrogen at the cathode and in practice some hydrogen production has to be accepted. This is of little consequence for a reinforced concrete section, but due to the fear of the possibility of hydrogen embrittlement of prestressing steel, this technique is rarely used on prestressed structures.

In most applications, the installation of anodes changes the appearance of the concrete and may leave unsightly patches. A solution to this problem is to paint the concrete as for the historic Thorverton Bridge, see Fig. 12.12.

Re-alkalisation

The reaction in cathodic protection is the basis of the patented technique of re-alkalisation, used to restore the alkalinity of carbonated concrete. In this technique, the reinforcement is connected as a cathode to a voltage source. A temporary anode is brought into contact with the surface of the concrete via an electrolyte, usually sodium carbonate. Under the impressed current, the cathodic reaction generates highly alkaline hydroxyl ions from oxygen and water. The hydroxyl ions repassivate the surface of the reinforcement. As with cathodic protection, several different types of anode may be used including steel mesh, titanium mesh (which is more resistant to corrosion) and electrolyte baths (for use on slab structures). Steel is more practical for re-alkalisation because the process is of shorter duration than chloride removal and less likely to consume significant quantities of the anode.

The voltages used in re-alkalisation are much greater than those used for cathodic protection being typically in the range 20 to 50 V, to give a current

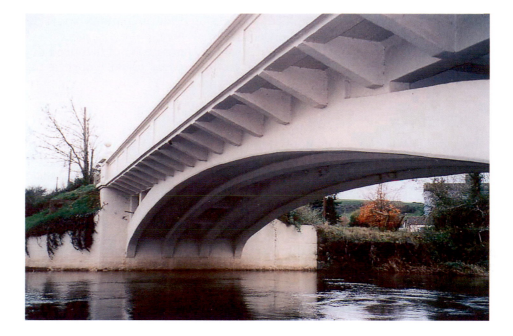

Fig. 12.12   Thorverton Bridge, Devon, 1908. Cathodic protection has been camouflaged by painting the concrete

density of 0.3 to 1.5 A/m². There is greater evolution of hydrogen and re-alkalisation is therefore unsuitable for prestressed structures.

Re-alkalisation usually takes between 3 and 6 days. It is claimed that the use of sodium carbonate as an electrolyte renders the concrete more resistant to further carbonation. Conversely it is claimed the highly alkaline sodium ions may trigger ASR in susceptible structures, and in these circumstances plain water may be preferred as the electrolyte. Re-alkalised concrete is said to be extra-resistive to carbonation, but it remains for this to be confirmed by years of experience and weathering in service.

Chloride removal

Also called desalination, this process is essentially the same as re-alkalisation except that water is used as the electrolyte. Negative chloride ions are attracted to an externally applied temporary anode and are thus removed from the concrete. Anodes are of the same form as used for re-alkalisation, namely mesh or liquid bath systems though desalination usually takes several weeks, rather than the several days typical of re-alkalisation. The desalination process can have the effect of increasing the permeability of the concrete and to offset this the surface may require impregnating with a suitable material applied after treatment.

The process can take 4 to 6 weeks to complete, but by no means all of the chlorides are removed. Concrete in the vicinity of the reinforcement is usually free of chlorides but further away, particularly behind the bars, the process is less effective. Typically, 50 to 90% of chlorides are removed overall.

With chloride removal, as with re-alkalisation, less than about 10 mm of cover can result in the reinforcement becoming an anode and preferentially corroding. In addition, all spalling, cracks and delaminations must be repaired or they will cause a short circuit.

The use of electrical methods in prestressed concrete is not favoured due to the fear that the hydrogen released in the process could embrittle the prestressing steel.

STRENGTHENING

Strengthening is most commonly required as a consequence of increased permissible vehicle weights and in cases when the structure has deteriorated to the extent that it has become weakened. Less commonly, mistakes made in design or construction may have to be rectified, but these are usually identified fairly soon after construction. Before strengthening is seriously investigated, advanced methods of structural analysis should be carried out to determine whether the bridge is indeed understrength. Methods of strengthening are described in the following sections and summarised in Table 12.5.

Identification of hidden strength

It is sometimes possible to raise the assessed load carrying capacity of bridges, and avoid unnecessary strengthening work or demolition, by identification of hidden strength. This can be regarded as a type of conservation. It is achieved by carrying out high level structural analysis to mobilise structural actions not recognised by the more commonly used and conservative assessment methods. On occasions it may be necessary to complement the analysis by strength tests on samples of the concrete and reinforcing steel to establish actual strengths as opposed to assumed values. Identification of hidden strength is a tool that can be used to enable bridges to remain in use and be conserved in their original state.

One of the most common examples of hidden strength is exemplified by the membrane action that is present in concrete slabs longitudinally and transversely restrained by abutments. This is sometimes referred to as flat arch action, a more descriptive term.

Plain slab decks are a fairly common form of construction for spans of up to about 7 m. They were designed to the standards then used and invariably had insufficient steel reinforcement to meet modern loading requirements. Nevertheless, hidden reserves of strength can often be identified. This was illustrated dramatically when TRL carried out load tests on the partially dismantled Wormbridge, near Hereford. The 380 mm thick slab deck had a clear span of 5.8 m and was reinforced with 305 mm by 127 mm rolled-steel joists, spaced 457 mm apart. There was steel mesh beneath the beams to hold together the concrete cover beneath the bottom flanges, but no transverse reinforcement. In a first test, 'patch' loading was applied to the deck at positions calculated to be most vulnerable using two 500 mm by 300 mm pads spaced 700 mm apart. A failure load of 2200 kN was recorded, compared with an assessed load, using conventional calculations, of 386 kN. In a second test, loading was via four 500 mm $\times$ 300 mm pads spaced 700 mm apart transversely and 1800 mm longitudinally. A failure load of 2904 kN was recorded, compared with an assessed load of 651 kN.

Table 12.5. Methods of strengthening

Method of strengthening	Advantages	Disadvantages
Identify hidden strength	Enables continued use without strengthening. Economic.	None.
Added structural elements	A structurally efficient method. Can be designed to have little or no impact on appearance.	The bridge loses its originality. Likely to disrupt traffic. Expensive.
Added reinforcement	Structurally efficient.	Likely to change appearance and require cosmetic overlay.
External reinforcement and sprayed concrete	Quick to apply. Relatively economic. Strong and dense concrete is ensured.	Changes appearance, usually unsightly. Only applicable to sound concrete. Requires skilful application. Expensive.
Bonded plating	No disruption to traffic. Economic. Quick to apply.	Changes appearance. Only applicable to good concrete, free of chloride. Susceptible to fire and impact damage. Requires periodic inspection to check for debonding.
External post-tensioning	Structurally efficient. Can be expensive, but usually cheaper than other options. Compressive stresses are introduced and the serviceability limit is raised.	Can change appearance. Requires careful design. Requires good corrosion protection.
Change structural action	Can increase both strength and durability. Can improve original design without change in appearance.	The bridge loses its originality. Can disrupt traffic. Expensive unless part of a wider range of measures.
Overslab	Minimal change in appearance.	Significant disruption to traffic. There are possible problems if utilities are present.

The Gifford method of compressive membrane analysis, the NLG program, was developed with the aid of scaled model tests and has been used successfully to reassess and upgrade the load carrying capacity of some eighteen slab decks.

There are occasions when it is helpful to supplement the analysis by carrying out load tests, in which case the Institution of Civil Engineers' *Guidelines on Supplementary Load Testing* (1998) should be followed.

Added structural elements

Construction of additional structural elements is a convenient method of strengthening suitable structures. Schemes can often be designed so that there is minimal change in appearance. The most common structural elements to be added are beams and columns.

Examples of bridges sensitively strengthened include Thorverton Bridge, across the River Exe in Devon, constructed in 1908, see Fig. 12.12. The structure has a single span of 25 m and was designed with reinforced concrete 'arch beams'. Over the years it deteriorated and was understrength to the extent that is was restricted to 7.5 tonne. A strengthening scheme was designed where two additional reinforced arch beams were installed. A thrust of 280 tonne was jacked into the new arch to reduce deformations and a new concrete deck installed. As the additional beams were located beneath the deck there is no change in appearance of the side elevation which therefore retains most of its originality. The work was completed in 1996.

Hampton Court Bridge across the River Thames in Surrey, is a three-span reinforced concrete arch structure built in 1931. When assessed it was found that there were local weaknesses in the deck slab spanning between columns and the arch. A strengthening scheme was designed having additional columns to reduce the length spanned by the deck slab. As the additional columns are behind the spandrel walls, there has been no change in the appearance of the bridge.

Added reinforcement

There are occasions when strengthening can be affected by adding reinforcing bars to the existing structure, for example to increase shear resistance. The bars can be installed in slots cut parallel to the concrete surface or holes drilled perpendicular to the surface. Added reinforcement is often carried out as part of a larger package of refurbishment work.

Old Bedford Bridge, Norfolk, is a three-span reinforced concrete beam-and-slab structure constructed in 1936. In the early 1990s it was found to be understrength and having spalled concrete and corroding reinforcement. Among other strengthening works, it was necessary to reinforce haunches in the longitudinal beams located by the intermediate supports. Slots were cut out by robot-controlled water-jetting and the reinforcement was fixed in place. The work was constrained by safety requirements and the location that was designated a site of special scientific interest (SSSI). On completion of the repairs, the concrete was re-alkalised and cosmetic coatings were applied to the concrete.

Woodbridge Old (so named to distinguish it from the new bridge beside it) across the River Wey in Guildford, Surrey, is a single-span, reinforced concrete structure having arch beams and constructed in 1912. When assessed in 1988, it was found that the beam-and-slab deck contained spalled areas, there was insufficient longitudinal steel in the top slab near the supports, and the stirrups were grossly overstressed. The shear resistance was increased by drilling centrally through the beams at 500 mm centres and grouting in 25 mm bars. In the event the drilling proved difficult as several of the cutting bits failed and it was necessary to switch to rock drilling. There were problems when reinforcement steel was encountered, the drill vibrated and sound concrete was broken off. The shear bars were grouted with proprietary non-shrink grout to provide the design pull-out strength within 150 mm of the bottom of the bar.

Bonded plating

In certain circumstance, it may be appropriate to strengthen an under-strength member by plate bonding. This technique consists of attaching steel or carbon fibre reinforced plastic plates to the external surface of the concrete using a suitable epoxy resin adhesive, as shown in Fig. 12.13.

In plate bonding, by adding additional reinforcement to the surface of the member, the ultimate strength of the section and its stiffness are increased. This technique can be used to increase the flexural strength and stiffness of slabs, beams and columns, and to a certain extent the shear strength of beams.

The increase in strength and stiffness achieved is critically dependent on the bond between the existing structure and the plates. It is essential there-fore, before strengthening a member by plate bonding, to establish that the

Fig. 12.13 Steel plating being bonded to bridge soffitt

risk of corrosion to reinforcement within the member is low and that the cover concrete, including any repairs, is sound and free from cracks or spalling. Before bonding, the concrete is grit blasted with a fine abrasive material to remove the surface laitance and expose the aggregate. Any small protrusions are removed by careful scabbling or grinding. When steel plates are used they should be sand blasted to remove all traces of rust and mill scale and cleaned with a suitable organic solvent to remove any oil or grease. After cleaning, the surfaces are immediately protected with a coat of suitable priming paint or aluminium spray to prevent corrosion. It should be noted that stainless steel plating is unsuitable, as it is difficult to obtain adequate adhesion.

Epoxy resins have been found to be the most suitable adhesives for plate bonding and can be applied either by trowel or by injection. The thickness of the adhesive coat, typically in the range 0.5 mm to 5.0 mm, has been found to have little effect on the strength of the connection.

The quality of the bond and hence the strength of the plated member is very sensitive to the skill and care of the labour used. Modern epoxy resins are supplied in 'two pack' form in the appropriate proportions for site mixing. Nevertheless, thorough mixing is required and it is essential that all of the hardener is used. It is also essential that all of the mating surfaces are adequately covered with adhesive and particular attention should be paid to the edges of the plate, which should be sealed with a resin putty or mortar to prevent the future ingress of water or other contaminants. Ends of the plates are bolted to the concrete to prevent debonding and peeling from the high stress concentration. The bolts which are grouted into the parent material beforehand, also aid positioning of the plates.

As the plates are unloaded when bonded to the concrete, they will not carry any permanent loads and are therefore effective for live loading only.

Bonded steel plating was introduced in the 1970s and numerous bridges have been strengthened, usually having the plates attached to their soffits. It has been found that mild corrosion can develop on the steel interface starting from the edges of the plates, although this has not caused any problems to date. Two pairs of bridges on the M5 motorway at Quinton were strengthened with bonded steel plates in 1975. The bonding has been monitored by impact-echo testing (hammer-tapping) and it has been shown that with the exception of a few small areas, it has remained generally sound. Initially 0.05% of the plated area was debonded, this increased to 1.5% after 18 years.

More recently carbon fibre reinforced plastic plates have become popular because they have a lighter weight, are not susceptible to corrosion, do not require bolting and are easier to use.

The method of plate bonding has the advantage that when applied to soffits, headroom is not significantly reduced and it is structurally efficient. However, carbon fibre reinforced resin plates can give off toxic fumes if exposed to fire.

All types of bonded plating require periodic inspection to check for debonding. Unless painted or otherwise disguised, bonded plating can change the appearance as shown in Fig. 12.14.

External post-tensioning

Significant increases in the strength of concrete members can be achieved through the use of retrofitted post-tensioning. Post-tensioning can be applied to ordinary reinforced or prestressed concrete. On smaller members it may be possible to apply the post-tensioning internally and this may be either bonded or unbonded. For larger members, external unbonded tendons are the only practical solution.

Retrofitted post-tensioning has the advantage that permanent stresses are imposed, cracks are closed and the serviceability limit state is raised.

When strengthening an existing structure with external post-tensioning, it is necessary to bear in mind that success will be dependent on the design of the anchorage and special consideration must be given to how the load from the tendons is transferred into the concrete at these locations. There are several possible methods of anchoring external post-tensioning, including reinforced concrete blisters, anchoring at existing diaphragms and anchoring at the ends of existing beams. In all cases, the concrete should be sound and free from chloride contamination. Cores should be removed from the concrete and tested to confirm whether its compressive strength is adequate.

Barrington Works Rail Viaduct, on a private railway in Cambridgeshire, is a reinforced concrete beam-and-slab structure having a 12.8 m river span and eight 7.3 m land spans, constructed in 1926. After 47 years without significant maintenance, it was necessary to strengthen the structure to carry 50 tonne wagons in place of the original design requirement of 34 tonne wagons. After a careful investigation of the state of the concrete, a post-tensioning scheme was designed. Steel anchorage brackets were fixed at each end and on either side of each beam using transverse Macalloy bars of 25 mm diameter stressed to 9 tonne and grouted. The longitudinal post-tensioning was by 25 mm Macalloy bars stressed to 8 tonne and locked-off

Fig. 12.14 Bonded plating. The appearance of a strengthened concrete soffitt

against the anchorage bracket. After removal of loose concrete and grit blasting of exposed reinforcement, galvanised mesh was wrapped around the sides of the beams and soffits. The assemblage was gunited to provide 38 mm cover to protect the Macalloy bars and brackets. As a result, the beams became somewhat bulbous in cross-section and disfigured. It was noted that there was a probable loss of about 25% of the gunite due to rebound and lack of hydration. The scheme avoided interruption to traffic and was economic.

Eraclea Bridge across the River Piave near Venice is a bowstring structure of 25 m span constructed in the 1950s. By the 1990s, it was evident that there was honeycombing in the concrete, transverse cracks in the deck, spalling and corroded reinforcement. One of the individual ties had failed several years earlier and been replaced, the others were strengthened. The bridge was being overloaded due to imposition of increased loading requirements since it was designed. Demolition and replacement was not considered an option due to unacceptable disruption to traffic and the historic value of the bridge. The deck was strengthened by the addition of 150 mm reinforced concrete, likewise the arch ribs. Each of the longitudinal tie beams was post-tensioned with 40 mm diameter Macalloy bars anchored in steel brackets designed to enable future inspection and located on either side of the beam. The total cost was comparable with a new structure, but remained the preferred option.

External reinforcement and sprayed concrete

The addition of external reinforcement, usually in the form of steel mesh, and sprayed concrete is a convenient method of dealing with understrength sections. It can also be used for strengthening masonry arch bridges as mentioned in Chapter 7. Sprayed concrete has advantages in situations where a reinforced concrete skin of 50 mm to 100 mm will suffice, because the method ensures the concrete is hard and dense. If necessary it can be applied in thicknesses of up to 300 mm.

On Axmouth Bridge, Devon, it was required to strengthen the substructure. Formed concrete was placed up to the level where the piers were above water or dry for six hours between tides. Galvanised mesh was fixed to the old concrete using stainless steel pins. Concrete was sprayed to a thickness of up to 100 mm to the height of the springing points of the arches. The concrete was composed of a mixture of Portland cement, zone 2 sand, stainless steel needles and a rapid hardener additive. The dry spraying process was used, the material being mixed with water at the gun head to produce a low water/cement ratio. This ensured a highly compacted and durable concrete.

Change structural action

Bridges can be strengthened by changing their structural action. Methods that have been used successfully include elimination of expansion joints to convert multispan superstructures from simply supported to continuous, and introduction of continuous reinforced concrete decks having independent support within the walls of arch structures. In both of these examples, there is no change in appearance.

Sand River Bridge, Free State, South Africa was due to be upgraded when it was damaged by flooding in 1988. It is a nine-span concrete arch structure having rubble infill. The flood damaged the deck and removed some of the infill rubble. In the reconstruction, the structural action was changed by having a continuous reinforced concrete deck integral with piers supported on the original piers to form a portal structure. Expansion joints were fitted at each end of the deck.

Mosel Bridge, Zeltingen, Germany is a seven-span reinforced concrete arch structure originally constructed in 1929. The three river spans were reconstructed in 1949 as three-hinge arches. In the late 1980s, the concrete had deteriorated and there was a need to widen and strengthen the bridge. As part of the works, and to improve the durability, the central hinges were closed and the river arches converted to a two-hinge action. This was achieved by casting an overslab continuous over the hinge. To avoid a reflection crack developing in the overslab, it was debonded from the original deck for 500 mm either side of the hinge.

Donner Summit Bridge, California, USA, an open spandrel arch bridge of total length 72 m and constructed in 1925, was constructed with five open expansion joints. After 70 years' service, the joints were badly deteriorated. As part of the strengthening work, the joints were removed and the bridge made continuous. Concrete on either side of the expansion joints was removed for a distance of 900 mm. The resulting gap was filled and an 85 mm thick reinforced overslab was cast in place.

Overslab

Overslabs are used on occasions when the concrete deck has become badly deteriorated and understrength, or is simply understrength for current load requirements. Depending on the design of the bridge, it may be possible to absorb the thickness of added reinforced concrete into the deck surfacing, otherwise there may be a step which has to be tapered by the approaches either side of the bridge. The taper has to be gradual to avoid development of impact loading as heavy vehicles run on to the deck. Overslabbing is economic and has the added advantage that waterproofing and drainage can be improved at the same time. It has the disadvantage that traffic is severely interrupted by the closure of at least one lane at a time while the work is carried out.

The deck of Woodbridge Old (see earlier discussion on added reinforcement) was found to have insufficient flexural strength. The deck was structurally sound, but there were numerous areas of spalled concrete. An overslab was cast on to the existing deck, structural connection being ensured by insertion and grouting of dowel bars at 500 mm spacing into the existing deck to act as shear connectors. The work was carried out in four phases to minimise disruption to traffic.

In Poland, testing and strengthening work was carried out on a four-span viaduct, constructed in 1951, and found to be understrength to carry modern traffic. The structure had a continuous deck designed to carry 15 tonne vehicles. Due to ineffective drainage, the concrete deteriorated over 40 years' service and chlorides caused the reinforcement to corrode. It was required to strengthen the viaduct to carry 30 tonne vehicles. Spalled concrete was removed from the upper surface of the slab and exposed reinforcement was cleaned and coated with epoxy. Longitudinal and transverse bars were fixed in place and a 120 mm concrete overslab was cast in place. The soffit was prepared in the same way, transverse reinforcement was fixed and covered in a 30 mm thickness of shotcrete. In addition to the added reinforcement, steel anchorage plates were bonded and bolted to the side faces of the deck and 16 mm diameter bars were fitted and post-tensioned transversely to the deck.

Donner Summit Bridge, California (mentioned in the previous section) was found to have deteriorated concrete and it was desired to strengthen the bridge to current vehicular loading requirements and seismic resistance. Among other works, the running surface and cover concrete on the upper face of the deck were removed. In places, concrete below the reinforcement steel had deteriorated due to de-icing salt and this also had to be removed. An 85 mm overslab, reinforced to strengthen the deck to carry full vehicular loading was cast in place. The overslab was designed to act as a horizontal diaphragm 'to tie the bridge together from abutment to abutment'.

MANAGEMENT PRACTICE

Assessment of load carrying capacity of older concrete bridges is invariably hampered by a lack of information and it is often necessary to check the sizes and spacing of reinforcement. Concrete cores may have to be removed for testing to determine actual compressive strength. Likewise, short lengths of reinforcing bar may have to be tested, but it is necessary to ensure that they are only removed from non-critical locations so that the bridge is not structurally weakened.

When bridges carrying normal traffic and exhibiting no signs of distress, are found to be apparently understrength, it is often appropriate to investigate whether there is any 'hidden strength'. This can be done by applying more advanced methods of analysis coupled by use of the results of available research as described earlier in this chapter.

Mass and reinforced concrete

As outlined in the previous sections, early concrete tends to have honeycombing and commonly has a low thickness of cover to the reinforced bars. There have been surprisingly few values of compressive strength reported for old concrete but those that have been evaluated are invariably high in relation to the default value (recommended for assessment) of 15 N/mm² and a value of 16.5 N/mm² generally assumed by designers up to 1917. For the bridges constructed up to 1940, protection of the reinforcing steel, afforded by the concrete cover, must be assumed to have been lost as the depth of carbonation will almost certainly have exceeded the thickness of cover. It follows that one of the key management tasks is to determine whether there is active corrosion using the methodology of inspection developed in recent years. For reinforced concrete, this involves the following:

- Visual inspection for cracking, rust stains, spalled concrete and exposed reinforcement
- Impact-echo testing (hammer-tapping) to detect hollow sounding areas where delamination has occurred and spalling is imminent
- Survey the thickness of concrete cover
- Measure the depth-profiles of chloride content and carbonation, and
- Survey electrode potentials to determine whether there is electrical activity.

The most likely places where reinforcement may corrode are:

- Where there is the least concrete cover
- Reinforcement located in corner positions, where there is concrete cover on two perpendicular faces.
- Stirrups (which tend to have low cover).
- At cracks that have formed in the concrete for one reason or another, permitting water, chlorides and oxygen to permeate more deeply into the concrete.
- Places where the concrete is subjected to leaking water, for example spillage from blocked drains, and failure points in the waterproofing on concrete decks and in the vicinity of expansion joints.

Expansion joints invariably leak for a large proportion of their lives permitting water and de-icing salt to drip on to the bearings, cross-beads and piers, causing corrosion. However, expansion joints were rarely used on the early concrete bridges so that they are generally free of this problem and it is more prevalent on structures built in the 1960s and later.

When corrosion is not identified but there are voided areas or cracks, it may be necessary to take corrective action. Voids can be grouted to strengthen the old concrete, provided there is a degree of interconnection between them. Cracks can be filled with a sealant as a protection against future problems. It is, however, essential to ensure that the causes of the cracking have been identified, as it is pointless to seal cracks that are progressively widening.

When corrosion is identified, it is necessary to remove defective concrete and clean the steel reinforcement, taking care to remove all corrosion products and concrete containing chlorides. The concrete must then be made good by patch repairs as outlined in the section on replacement of defective cover. Care should be taken to ensure that the repair mortar is compatible with the old concrete in all respects, including appearance.

On balance, painting to camouflage unsightly repairs is not recommended, except in circumstances that especially merit it, as paint has a finite life and can become exceedingly unsightly towards the end of its life. Reflective cracking can develop in the paint soon after application. Also, painting detracts from the originality and attractiveness of old weathered concrete. However, colourless penetrant coatings provide protection against ingress of contaminants, such as chlorides, without these disadvantages.

Application of sprayed concrete has many advantages, for example good adhesion and high density. However, it is not recommended for use on heritage structures in locations that can be seen, as its appearance is not compatible with old and weathered concrete.

Electro-chemical methods can be used for treating corrosion:

- Cathodic protection halts active corrosion
- Chloride removal deals with the root cause of corrosion, and
- Re-alkalisation restores the protective properties of the concrete.

The pros and cons of electrochemical methods are summarised in Table 12.4

Prestressed concrete

The above applies mainly to reinforced concrete, which comprises the majority of concrete bridges constructed before 1960. There are, however, a small number of early prestressed and post-tensioned bridges and these present special problems, particularly in relation to inspection. The collapse of Ynys-y-Gwas Bridge in 1985 and the subsequent failure investigation (see 'Performance of concrete') showed that corrosion of the post-tensioning steel can progress to a very advanced stage without exhibiting any discernible evidence on the external surfaces of the concrete. Although considerable research has been carried out to develop non-destructive and semi-destructive methods of inspection, the most effective and reliable method is to physically expose the steel. This is achieved by identifying as accurately as possible, the location of the post-tensioning ducts and drilling, or coring, a hole close to the depth of the ducting. The ducting is then gently exposed and peeled back. The inside can then be inspected to determine whether it has been properly grouted; if not, the steel tendons will be exposed as shown in Fig. 12.15, so that an endoscope can be inserted into the duct and any corrosion will be evident. If the duct has been well grouted, there is a low likelihood of corrosion, but it is prudent to gently remove the

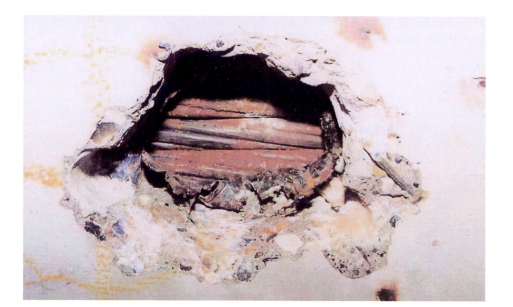

Fig. 12.15 Invasive investigation of duct. Although there is no grout in this duct, the strands are in good condition having only light surface rust

grout and expose the steel. There have been occasions when severe local corrosion has occurred beneath the grout. This can be due to chlorides penetrating through the grout or being present in the original mix, as calcium chloride, to accelerate the curing process.

The prestressing tendons are most likely to corrode at places where voids can form during the grouting process, leaving the steel unprotected at:

- high points in the curvature of the ducts, and
- ends of ducts in the vicinity of anchorages.

Corrosion can occur, less commonly, at places where the tendons touch steel components and the grouting is without significant voids but may contain chlorides. Likely locations are:

- positions where the tendons touch the steel duct, and
- positions where the tendons touch steel separators (separators were used to keep wires apart in some of the earlier prestressing systems)

Corrosion can also occur at places in the vicinity of connections between precast post-tensioned concrete segments, particularly when dry packed mortar was used to seal the connection. Bridges particularly at risk are those having running surfaces laid directly on to the prestressed beams.

When ducts are found to be empty and the steel tendons (wires, strands or bars) are only lightly corroded, or free of corrosion altogether, it may be appropriate to regrout the ducts. This is explained in the section on conservation of prestressed concrete.

When ducts are found to contain voids and the tendons exhibit a degree of corrosion, there is a risk that the situation may worsen. In this case, and depending on the circumstances, it may be appropriate to monitor the post-tensioning rather than take immediate action. Monitoring can be limited to simply inspecting the concrete for the development of cracks and evidence of corrosion activity, such as staining on the surface of the concrete caused by corrosion products. More sophisticated monitoring can be carried out using instrumentation such as acoustic emission to detect the release of energy when a wire fractures. Acoustic emission has been successfully used in the field and can be set up and automated to transmit data to the bridge engineer's office.

Monitoring may also be carried out to 'buy time' when corrosion is more advanced, but it is not appropriate to take immediate action.

When the post-tensioning steel is found to be seriously corroded, it invariably affects a significant number of the tendons and some are likely to be fractured. Under these circumstances, decks have usually been demolished and replaced. When it is considered appropriate to repair them, methods such as external post-tensioning or bonded plating, may be feasible, as described in the section on strengthening.

An example of a post-tensioned segmental footbridge that has not required maintenance and has weathered well is shown in Fig. 12.16.

Table 12.6. Examples of concrete bridges

Bridge	Engineer	Date of construction	Span(s) (m)	Comments
Flixton Estate, Homersfield, Suffolk	T and N Phillips	1870	16.5	Single-span, riveted iron frame encased in concrete. Refurbished 1993, grade II listed. Originally a private bridge, now pedestrianised.
Axmouth, Devon	Phillip Brannon	1877	9.1–15.2–9.1	Three-span mass concrete cellular arches. Refurbished 1989, ancient monument now pedestrianised.
Glenfinnan Viaduct, Scotland	Simpson and Wilson	1897	15m × 21	21-span mass concrete rail viaduct, still in use.
Chewton Glen, Hants		1901	5.5	Reinforced concrete arch clad by brick spandrels. Grade II listed, but arguably not a good example of reinforced concrete.
Royal Tweed, Berwick	Mouchel and Partners	1928	57–75–86–110	Four-span open spandrel arches. Refurbished 1980.
Waterloo Bridge, London	Rendel, Palmer and Tritton	1944	76m × 5	Five-span, commenced 1937 and completed 1944. Carries full traffic loading, no significant maintenance required.
Nunn's Bridge, Fishtoft, Lincs	Mouchel and Partners. Prestressed Concrete Ltd	1948	22.5	Single-span post-tensioned beam and slab. No significant maintenance, still in use.
Rhinefield, Hants	E W H Gifford, Hants CC	1950	10	Single-span post-tensioned. No significant maintenance, still in use.

Fig. 12.16 Bradford-on-Avon footbridge, 1962. A post-tensioned structure having precast segments. The concrete has weathered in sympathy with the surrounding stone walls and old buildings

BIBLIOGRAPHY

Mallett, G.P. (1994) *Repair of Concrete Bridges*. TRL state-of-the-art review, Thomas Telford, London.

Broomfield, J.P. (1997) *Corrosion of Steel in Concrete,* Spon, London.

Hampshire County Council. (2000) *Bridges in Hampshire of Historic Interest*, Hampshire County Council, Winchester.

Chrimes, M.A. (1996) *The Development of Concrete Bridges in the British Isles prior to 1940*. Proc ICE, Structure and Buildings.

Public Works Roads and Transport Congress. (1933) *British Bridges.*

The Engineers' Year Book. (1923) Crosby Lockwood and Son.

Brown, J.H. (1987) *The Performance of Concrete in Practice: A Field Study of Highway Bridges.* TRL Contractor Report 43, Crowthorne.

The Institution of Civil Engineers. (1998) *Supplementary Load Testing of Bridges,* Thomas Telford, London.

Wallbank, E.J. (1989) *The Performance of Concrete in Bridges – A Survey of 200 Highway Bridges.* London: HMSO.

Andrew, A.E. and Turner, F.H. (1985) *Post-tensioning Systems for Concrete in the UK: 1940–1985.* CIRIA Report 106 Construction Industry Research and Information Association, London.

Chapter 13

The architecture of bridges

GLOSSARY

Arrises	The edges of metal, timber, brick or stone components
Ashlar	Smoothly dressed (faced) and squared stones laid to precise horizontal courses
Classical	Architectural style derived from ancient Greek and Roman architecture
Corework	Material in the centre of masonry construction, often poor quality infill between better-built facework
Corinthian	A Greek order of architecture using columns, the capitals of which are carved in the form of Acanthus leaves
Diaper work	Diagonal pattern, usually of differently coloured bricks, often achieved by incorporation of dark over-burnt headers
Doric	A Greek order of architecture using columns with simple moulded capitals and often no base
Facework	The visible outer part of masonry construction, often better built than the interior 'corework'
Garlanded	Architectural decoration of fruit or foliage in circular form
Guano	Faeces and accumulation of other debris left by pigeons and other birds

Guttae	Literally 'tears', or small dowel-shaped projections at the bottom of triglyphs
Ionic	A Greek order of architecture using columns, the capitals of which are carved with large spiral-shaped volutes which copy the shape of cochlea sea shells or their larger fossilised 'ammonite' predecessors
'Mathematical' bridges	These are designed, in theory, according to mathematical principles, so that loads are taken by pure axial compression in each structural member down to the foundations
Modillions	Rectangular projections in the upper part of a classical frieze which imitate rafter-ends of a timber roof
Neo-classical	Refinement of classical style introduced into England during the latter part of the eighteenth century
Palladian	A classical style of architecture following rules set by the Italian architect Palladio, introduced into England in the seventeenth century
Pediments	Classical triangular decoration, usually at the top of an architectural feature (door, window, etc.). Derived from the shape of the end of a roof
Quoins	Corner stones or bricks often projecting beyond the general wall face, when they would be 'rusticated' (q.v.).
Refuges	Small increases in the width of a bridge, usually over the cutwaters, where pedestrians can take 'refuge' from passing traffic
'Ribbon' pointing	Pointing that projects beyond the face of masonry, usually regarded as bad practice.
Rough random	Rough-faced masonry laid to irregular pattern with no horizontal courses.
'Roving' bridges	Canal bridges that allow a horse towing a barge to cross from one bank to the other without untying or entangling the tow-rope.
'Rubbers'	In brickwork, soft bricks (usually red) that can be 'rubbed' or shaped, often to form a flat brick arch
Rusticated	Stones emphasized as being separate by having chamfered edges
Skewed masonry	Bricks or stones laid to cross-sloping courses because the arch of a bridge is not at right-angles to the length of the bridge

| Torchères | Supports for free-standing light fittings, often decorative cast-iron |
| Triglyphs | Shallow rectangular projections in the centre of a classical frieze, with vertical grooves, which imitate beams ends in timber construction |

BACKGROUND

A bridge, almost by definition, indicates a structure of engineering competence, but what is its architectural component? Certainly, it is not merely the decorative trimmings that may be applied, though they may contribute to the overall architectural effect. Rather, it is related to the aesthetic appearance of the design, as a unit of construction, and to the impact of that design on its surroundings.

Bridges have evolved, like other buildings, through various historical styles. Thus the Greek 'beam and slab' form of construction, seen initially in the simplest of buildings and in the early 'clapper' bridges, has developed through timber structures to modern steel and concrete beam and slab bridges. Likewise, introduction by the Romans of the arch-form, seen in their aqueducts for example, set the scene for development of many major arched bridges. Suspension and cantilever bridges, perhaps, have less architectural precedent, their forms being found in nature, for instance in spiders' webs, vines and in tree branches.

Engineering criteria for bridge construction include the width of obstacle to be spanned, nature of the ground and abutments and the effect of environmental conditions on the structure. Architectural criteria, almost conversely, include the overall impact of the structure on the environment and the unifying of the various elements of the structure into an aesthetically pleasing whole.

A 'pure' structure may display no overtly architectural trimmings, yet be acknowledged as having fine architectural qualities. Brunel's Tamar Bridge and Baker and Fowler's Forth Railway bridge, also some relatively modern award-winning concrete bridges come within this category. In other cases, careful application of non-structural architectural features has transformed an otherwise utilitarian engineering solution into a pleasing overall design.

ARCHITECTURAL TREATMENT GENERALLY

Many architecturally decorative features have evolved from forms found in nature, for instance the 'Corinthian' capital from the acanthus leaf or the 'Ionic' volute from the cochlea or ammonite. Other features replicate or stylise constructional details. For example, in the Greek classical orders, 'modillions' represent rafter ends, 'triglyphs' beam ends and the 'guttae' their securing dowels. In bridges, as in other structures, application of architectural decoration is often most appropriate where it fulfils, or appears to fulfil, a function related to the derivation of that detail. Decoration applied in other circumstances is more likely to be superficial and, being unrelated to the structure, less likely to contribute to an overall pleasing design.

SCULPTURE ON BRIDGES

Sculpture on bridges has been important at many times. The Roman Ponte d'Angelo of 136 AD has Bernini's famous angel statues added in 1668. The fourteenth century Karlsbrucke in Prague is famous for its thirty baroque statues of 1707. Around 1900, they were popular all over the world, for example the bronze winged horses on Arlington Memorial Bridge, Washington, (1932) shown in Fig. 13.1. Repair and conservation of statuary is a specialist's task, but patinated bronze should not be cleaned or polished with anything. Gilded bronze can be regilded, but not mechanically burnished or painted. Terracotta (e.g. the Westminster Bridge lion) should only be washed with water and never blasted with any abrasive, since this would remove the delicate weather resistant fireskin. The cleaning of stone varies considerably, dependent on the type of stone and type of dirt. Restoring concrete sculpture is a specialist version of patch repairs to concrete structures, e.g. Taft Bridge, Washington (1907). The repainting of cast iron sculptures (e.g. Vauxhall Bridge, Fig. 13.4) should follow similar guidance to that on cast iron structures.

Fig. 13.1   Arlington Memorial Bridge, Washington, USA

BRIDGE DESIGNERS

Although with the advent of the Industrial Revolution engineers were increasingly commissioned as bridge designers, before the beginning of the nineteenth century most of the leading architects had designed major bridges, especially in architecturally or environmentally sensitive locations. Up to the present day, architects have worked in partnership with engineers at the leading edge of bridge design.

The eighteenth century landscaped park almost always included an artificially created lake and a classical bridge. Vanbrugh's grand design for Blenheim Park bridge (1711) was never completed, yet it contains a small theatre within its abutment and is itself like a piece of scenery for one of Vanbrugh's stage sets. Robert Adam designed the delightful bridge, Fig. 13.2, between two lakes at Kedleston Hall, Derbyshire and James Paine built another at nearby Chatsworth (1762). William Kent designed the Palladian bridge with colonnades and pediments at Stowe (1734), whilst also designing the gardens and mansion.

Other Palladian bridges followed at Wilton Park, Wiltshire (1737), and at Prior Park, Somerset (1750), Fig. 4.8. Sir William Chambers' heavily rusticated bridge at Woburn Abbey, Bedfordshire (1760), Fig. 13.3, is one-sided, since there is a weir almost level with the roadway on the upstream side. Sir John Soane designed a beautiful single-arched bridge at Tyringham, Buckinghamshire in 1797.

From the end of the seventeenth century onwards, with stone-arched

Fig. 13.2 Kedleston Hall Bridge, Derbyshire

Fig. 13.3   Woburn Abbey bridge, before repair

bridges being a known technology, architects were frequently engaged to assist in the design of major bridges.

Thus Nicholas Hawksmoor designed St John's Bridge, Cambridge (1698) and Scottish architect William Adam designed a remarkable five-arched bridge at Aberfeldy for General Wade (1733). Sir Robert Taylor designed the Thames bridges at Swinford, Oxfordshire (1769) and Maidenhead (1772). At the same time, John Gwynn designed Magdalene Bridge, Oxford, followed by three others over the river Severn at Atcham, Shrewsbury and Worcester.

The well-known iron bridge at Ironbridge, Shrewsbury, was designed by the architect Thomas Pritchard in 1779, working in collaboration with iron-masters Abraham Darby and John Wilkinson. Pritchard had earlier been responsible for other bridges made with stone and iron.

Thomas Harrison designed bridges at Derby and Lancaster, then built what is still the longest single-span masonry arch in England, 61 m across the River Dee at Chester (1827), Fig. 4.6. Returning to the Thames, John Haywood was engaged at Henley (1782), Sir Jeffrey Wyatville at Windsor (1824, concurrently with rebuilding at Windsor Castle), and Sir Charles Barry at Westminster (1860, in conjunction with reconstruction of the Houses of Parliament). Sir Horace Jones added some architectural splendours to London's Tower Bridge (1894) and Richard Norman Shaw added architectural interest to Vauxhall Bridge (1906), Fig. 13.4. This was paralleled abroad by Otto Wagner at Ferdinandsbrucke in Vienna (1900) and by Alby at Pont Alexandre III in Paris (1900).

From the end of the eighteenth century, with the increasing availability of iron, a new breed of technology developed. Thomas Telford (1757–1834) was the first President of the Institution of Civil Engineers. He developed

Fig. 13.4   Vauxhall Bridge, London

the use of iron, especially in suspension bridges, although he began working life as a mason's apprentice and designed the magnificent Dean Valley Bridge in Edinburgh. John Rennie (1767–1821), on the other hand, designed many fine masonry bridges. These engineers, together with James Brindley also designed many canal bridges and aqueducts, whilst the principal railway bridge builders were Isambard Kingdom Brunel (1806–1859) and the Stephensons, George (1781–1848) and Robert (1803–1859).

In the second quarter of the twentieth century, there was a demand for vast numbers of medium-sized bridges for the growth in the road network in Britain, and much of Europe. This happened to coincide with a period where the technology for such bridges was well known, so there was an enormous use of architects to embellish designs produced by engineers. The Thames bridges of this period involve a roll call of the most famous architects of the era: Runnymede, 1939 (built 1961), Edwin Lutyens; Hampton Court, 1933 (Fig. 13.5), Edwin Lutyens; Chiswick, 1932, Sir Herbert Baker; Lambeth, 1932, Sir Reginald Bloomfield; Waterloo, 1944, Sir Giles Gilbert Scott; Southwark, 1921, Sir Ernest George.

The best and the most astonishing of the European designers was Piet Kramer, who designed over 200 bridges in and around Amsterdam from 1917 until the 1950s. The bridges, though small, ripple with wonderful detail in brick, stone and steel. The craftsmanship is exquisite and the forms user-friendly and beautiful. Many have witty sculpture, such as sea monsters or businessmen on the phone, Fig. 13.6.

Fig. 13.5   Hampton
Court bridge

Sir Edwin Lutyens was without doubt the best of the British bridge architects of the period, and also the most prolific. Apart from his Thames
bridges, others are the bridges in Surrey, designed with W.P. Robinson the
County Engineer, including Pilgrims Way Bridge at Guildford, the Mole

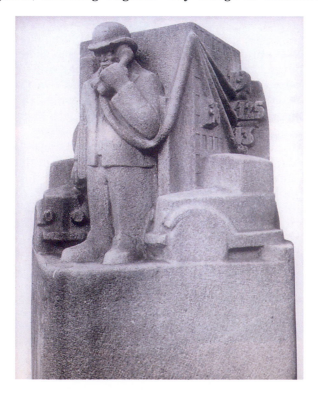

Fig. 13.6   Galenstraat
bridge, Amsterdam

Bridge at East Molesey and a dozen bridges along the A30 Staines bypass. They tend to use local brick with stone trimming, all beautifully detailed, as good as any of his great country houses. The proportions are elegant and graceful. The form and details are appropriate to the location: grand and florid near a palace, but simple and rustic in the countryside. The quantity and quality of the detail reflects the proximity of the viewer, particularly for pedestrians and when viewed from a boat. However, he is very much an engineer's architect, he is always careful to express truthfully the structure for what it is, for example a concrete arch is expressed as such on the facade, with the brickwork carefully shown to be non-load-bearing. Sadly many of his bridges are now unappreciated and not maintained with the care with which they were designed.

LAMPS AND LIGHTING

Although, nowadays, roads are illuminated in rural as well as in built-up areas, lighting on historic bridges was almost solely confined to the urban environment.

Clearly, the older town bridges would have had no lighting, but sometimes this has been added with more or less sensitivity. Bideford Bridge, dating

Fig. 13.7 Wheeling Bridge, West Virginia, USA

from the fifteenth century, but last widened on each side in 1925, now has iron torchères at regular intervals supporting lanterns of that period, though now containing fluorescent lamps.

Nearby at Barnstaple, however, a similar multi-arched bridge of medieval origin built across the estuary has a forest of much less-elegant conventional tall lighting standards with sodium lamps. The standards bear no relationship to the bridge in scale or appearance.

It is important that where possible due recognition should be paid to the scale and materials of an historic bridge, whilst maintaining the required level of illumination. Access for maintenance must also be borne in mind. This has been achieved quite harmoniously in many places, for instance at Bradford-on-Avon, where elegant iron standards have been attached around the stone parapet balustrades. The lanterns illustrated are of modern design (a variant of the 'Westminster' lantern), and are fitted with high-efficiency lamps, yet are in scale with the older standards on which they are mounted.

Unfortunately, modern lighting on old bridges is not always treated so sympathetically. The otherwise attractive old suspension bridge at Wheeling, WV, USA, Fig. 13.7, is marred by incongruously modern light fittings.

Where older lanterns survive on historic bridges they should be retained and kept in good repair wherever possible. The example illustration at Dur-

Fig. 13.8 Frammel Gate bridge, Durham

ham, Fig. 13.8, complete with its ventilated capping, still retains its original gas gear, although it appears to have been converted to electricity.

This 'Grosvenor'-type lantern has the more expensive circular glass enclosure, but the more commonly found 'Windsor' variant is shown in Fig. 13.9 with four flat sides. Here, at Queens Avenue Bridge, Aldershot the lantern is mounted on an exceptionally splendid cast-iron torchère standard and plinth, which relates well in scale with the substantial stone-capped brick pier supporting the whole ensemble.

The 'classic globe' lantern, on a cast iron post, stands well on the broad parapet of Axmouth Bridge, Fig. 13.10, a popular type also found at Shoreham and in many other places.

The London Thames bridges, unsurprisingly, have some of the best examples of torchère lighting, as shown in Figs 13.11 to 13.13. All are designed to be appropriate to their context. Thus the triple lanterns on Westminster Bridge have crocketted brackets and gothic-styled cast-iron supports echoing the architecture of the Palace of Westminster, both being designed by Sir Charles Barry, who engaged A.W.G. Pugin to produce detailed drawings. Lambeth Bridge has more severely designed neo-classical torchères very typical of Blomfield's work. Twickenham Bridge is lit by elaborately designed bronze lanterns, quite a rarity! Truly splendid art deco lamp standards for Chelsea Bridge were designed by the noted sculptor

Fig. 13.9 Queens Avenue, Aldershot

Fig. 13.10 Axmouth
Bridge, Devon

Joseph Armitage, who also designed the well-known emblem for the National Trust.

Another aspect of bridge lighting is illumination of the structure itself. This may be achieved by floodlighting or by highlighting the profile of the structure. Professional advice by specialist lighting designers is necessary for both types of illumination. Many of the Thames bridges are lit in one or both of these ways. The masonry or composite bridges lend themselves best for floodlighting, for example Tower Bridge and Waterloo Bridge. The latter, with its light-coloured shallow arches, responds well to up-lighting below each arch. Obviously there is less substance in suspension bridges so the Albert Bridge, for example, is delineated very well at night by its profile picked out in hundreds of individual lamps. Another, but much larger and more modern suspension bridge, the Pont du Normandy near Le Havre in France, is spectacularly lit by moving beams of laser light.

Provided that light sources do not dazzle users on or below a bridge, there is considerable scope for effective and efficient illumination of historic bridges of all kinds, but the luminaries themselves should not be obtrusive during the daytime.

Regard must be paid when considering illumination, to any statutory requirements, for instance traffic lights, railway signals and navigation lights. Whilst if they are necessary, they should be discreetly, but clearly located on a historic structure, other forms of lighting must not be allowed to distract bridge users from the primary purpose of any statutory lighting.

Fig. 13.11 Westminster Bridge, London

PARAPETS AND BALUSTRADES

The earliest clapper bridges have no parapets, but they usually span very shallow watercourses. Some early arched bridges for instance the 'Roman' bridge on the island of Torcello in the Venetian lagoon, also have no side protection, it being difficult to construct a secure parapet at the edge of a very thin arch. Many railway bridges, also, were built without parapets, they being unnecessary for the protection of passing trains. However, nowadays for the safety of railway maintenance gangs some form of barrier is being required.

The early packhorse bridges (e.g. Anstey, Leicestershire) had very low parapets, little more than high kerbs to guide the traveller away from the edge on a foggy night. Where remaining, these are almost always now only in pedestrian use. Their elevation above streams is usually very modest and it is to be hoped that strict application of safety legislation will not require disfigurement of these historic bridges.

Where the rise over arched masonry bridges became greater, solid stone parapets became the norm. Being increasingly vulnerable to damage from passing traffic, many parapets have been reconstructed, not always according to the designer's original intentions. For instance the fine New Bridge (c1520) across the river Tamar at Gunnislake (said by Pevsner to be

Fig. 13.12   Chelsea
Bridge, London

Fig. 13.13   Battersea
Bridge, London

Cornwall's finest bridge, Fig. 13.14) is constructed in granite ashlar except for the parapet, which has been reconstructed in rough random coursed granite. Whilst raising or strengthening masonry parapets may be necessary on an historic bridge, this should always be carried out in sympathy with the original structure, concealing any necessary reinforcement.

Sometimes when a bridge has been widened, the opportunity has been taken to raise or strengthen the parapet. Unusually when the fifteenth century Rothern Bridge at Great Torrington, Devon (Fig. 13.15), was widened, arches and parapets were formed all in vertically coursed granite, an extremely strong method of construction.

For reasons of economy, bridges were usually built as narrow as possible, even though many were subsequently widened. On long multi-arched bridges, where traffic was heavy, this could be hazardous to foot travellers if they were unable to step aside from passing carts and coaches. Pedestrian 'refuges' are therefore commonly found conveniently built over cutwaters, a distinctive feature of many masonry bridges. In country areas, where there are no footpaths across bridges, these refuges remain a useful haven from increasingly heavy and fast road vehicles. On railways, also, it is a requirement for refuges to be provided at regular intervals on bridges, as well as in cuttings and tunnels.

As the design of masonry bridges developed in the seventeenth and eighteenth centuries, parapets became more elaborate. Clare Bridge, Cambridge, built in 1644, is adorned with a balustraded parapet enriched by carved panels over the cutwaters and keystones, surmounted by fourteen stone balls. Hawksmoor's bridge for St John's College, Cambridge (1698) has

Fig. 13.14 New Bridge Gunnislake, Cornwall

Fig. 13.15 Great Torrington bridge, Devon

enriched balustraded parapets of similar design, but without the stone balls. Vanbrugh, at Blenheim in the early eighteenth century, integrated the parapets with the extended voussoirs from the main arch in a typically baroque gesture.

Balustrades, rather than solid parapets, became the norm in the eighteenth century. John Gwynn's Magdalen Bridge at Oxford, for example, has balusters interspersed with plain ashlar die blocks and projecting piers over the cutwaters. This balustrade has been repaired in recent years, badly decayed balusters being renewed in matching stone.

Canal bridges almost invariably have stout parapets, often in engineering brickwork with heavy half-round granite copings. Because these hump-backed bridges in country areas are often on busy narrow country lanes, their parapets tend to suffer from vehicles scraping the sides or, even worse, demolishing the brickwork. Protective bollards and kerbs help to preserve such parapets where road conditions permit. It is interesting to note that, whilst brickwork on the face of canal bridges is laid in horizontal courses, that to the parapets follows the slope of the pathway thus avoiding cut-bricks. The most elegant canal bridge parapets are found on 'roving' bridges where the canal path (and the parapet) rises on one side of the canal bank, passes over the bridge and drops to the other bank.

Decorative brickwork is less often found in bridge parapets although, as noted above, Sir Edwin Lutyens was skilled in its use. Parapets on the little bridge at Elvetham Hall by Teulon are good examples of this type of brick-work. The patterns are produced from relatively soft red brick 'rubbers' which are sometimes prone to erosion, which is exacerbated if a hard mortar is used when repointing is carried out. Replacement bricks should be carefully matched to the original brickwork and only a 'soft' lime mortar or lime putty should be used for bedding and pointing.

Timber parapet rails are not very common because maintenance can be heavy but there are some notable exceptions. The 'Chinese' footbridge at Pain's Hill, Bath, is an elegant example and others are the so-called 'Mathematical' bridges at Cambridge and at Whitwick, West Midlands.

With the advent of iron bridges came iron parapets or railings. The simplest iron railings are found on the earlier suspension bridges, for instance Brunel's bridge (finished after his death), over the Avon Gorge at Clifton, Bristol. A simple and effective parapet is one of woven wrought iron bands, often with a cast iron capping. This was a favourite device on the many delightful railway footbridges still surviving. Wrought iron needs to be kept well painted, with good maintenance, or corrosion can build up in the many lapped joints.

Cast iron is less prone to corrosion and has been used for parapets of many of the Thames bridges, for instance in Westminster and Lambeth bridges, where its ability to be cast in decorative shapes is seen to good advantage. Twickenham Bridge has parapets of concrete and bronze, which is relatively corrosion-resistant. Waterloo Bridge has steel parapet rails set directly into a Portland stone plinth. Unfortunately there seems to be no provision for thermal movement and the stone has fractured in many places.

The early concrete bridges built up to about 1940 were designed with decorative features, for example panelling, balustrades, pilasters, etc., Fig. 13.16, and their appearance is more interesting than the minimalist designs that came later.

Fig. 13.16 Fen Causeway, Cambridge

SURFACE DECORATION

Many masonry bridges are faced in roughly coursed stonework that collects lichens and merges well with their rural settings. Surface texture is important in large-scale structures like bridges, as also their detailing so that the structure reads well when viewed both from a distance and close at hand. Thus some of the finest viaducts are built in massive roughly rusticated blocks of stone, but with their arrises finely dressed. It is important, when repair or renewal of masonry is required, that the scale of large blocks is maintained, with the detailing repeated in any new work.

Surface decoration of parkland and urban bridges is usually more sophisticated, with classical or gothic vocabulary according to the style of adjacent buildings. In the seventeenth and eighteenth centuries classical styles were almost universal, with the return of gothic architecture from the early nineteenth century. Medieval gothic bridges, almost universally, have very restrained surface decoration, most of the mouldings being derived from a structural or constructional requirement. In some ways, this kind of embellishment might be said to be the most 'true to form'.

Decoration arising naturally from the structure includes moulded ribs to the arches of medieval bridges, Fig. 13.17. Projecting band courses or corbels, at the springing of arches were often provided to support centring for the arches, with putlog holes left as decorative features when the centring and access scaffolding was removed.

If cutwaters did not rise to provide pedestrian refuges, the offsets were sometimes decoratively treated, the decoration ranging from stepped offset stones to leaping dolphins on the English bridge at Shrewsbury and larger

Fig. 13.17 Greystone bridge, Bradstone, Devon

than life-size bronzed figures (of the arts, sciences, etc.) at Vauxhall Bridge, London, Fig. 13.4.

The base course of a parapet is often marked by a projecting moulded 'string', sometimes this delineating the change from horizontal courses in the walling below to sloping courses of the parapet, see Fig. 13.17.

In the seventeenth and eighteenth centuries, the scope for surface decoration of classical bridges was extended considerably. Abutments, plinths, quoins and arch voussoirs were often emphasised by rustication (bold v-jointing of stones). Hawksmoor's St John's Bridge at Cambridge, unlike the similar but earlier Clare Bridge, has boldly rusticated voussoirs extended to fill the whole of the spandrels between the cutwaters. Vanbrugh, at Blenheim, carried this idea even further, as noted in the previous section on parapets.

Hawksmoor and subsequent architects also employed a comprehensive range of classical detailing to parapet string courses and copings, especially when these were able to be viewed close at hand. It is interesting that Vanbrugh's unfinished design at Blenheim, although displaying dramatic use of classical motifs, is completely lacking in detailed mouldings. This is because the bridge was designed primarily to be seen as a distant view across the park.

The Palladian bridges at Stowe, Prior Park (Fig. 4.8) and Wilton Park, on the other hand were designed to be enjoyed, not only from a distance, but when crossed as part of a tour of the grounds. They are thus full of intimate

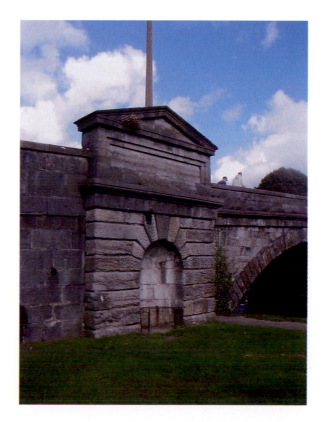

Fig. 13.18 Totnes
bridge, Devon

classical detail. In bridges of this kind the detailed mouldings need to be understood and respected in any repairs.

In the previous section, decoration on parapets has been noted, but arch spandrels and abutments also offered scope for decorative treatment. The spandrels at Kedleston Hall bridge as in Fig. 13.2, for example, are adorned with low-relief rosettes very typical of Robert Adam and more often recognised in his ceiling designs. Kedleston, Chester and Totnes bridges also have elegant semi-circular niches in their abutments as in Fig. 13.18. In the latter two, the abutments are designed as classical pavilions, with pediments set over friezes, Chester is complete with triglyphs and guttae.

The architectural desire to express the importance of a bridge and give an otherwise utilitarian structure some presence was sometimes achieved by the construction of ornate pavilions. Examples include the Balcombe Rail Viaduct in Sussex and Rochester Bridge, shown in Fig. 5.19.

Engineer John Rennie was fond of decorating his bridges with classical pilasters, or columns, as at Kelso bridge, Roxburgh. These serve no very practical purpose, but add stature to what would otherwise be rather plain structures.

As with the Palladian bridges, scope for surface decoration increases when the bridge structure rises above parapet level. This is particularly so

Fig. 13.19 Menai Strait's Bridge

Fig. 13.20   Clifton Suspension Bridge

Fig. 13.21   Albert Bridge, London

with 'covered' bridges. Thomas Rickman's gothic Bridge of Sighs at St John's College, Cambridge and the classical buildings on Pultney Bridge, Bath, being good examples. Suspension bridge towers also provided scope for architectural embellishment: Telford's Menai Strait's Bridge, Fig. 13.19, and Brunel's Clifton Suspension Bridge towers, Fig. 13.20, are especially good examples.

Marlow Suspension Bridge and the Albert Suspension Bridge, Fig. 13.21 both have decorative iron suspension towers.

As already noted in parapets, cast iron gave designers considerable freedom for enriching their bridges with surface decoration. Perhaps the most spectacular decorative use of this material is seen in the spandrels of Telford's bridge over the river Conway at Betws-y-Coed. A riot of roses, thistles, shamrocks and leeks fills the spaces over arched inscription borders. The North Road Bridge, Exeter, is another splendid example of cast iron decoration, Fig. 13.22.

Other bridge engineers added architectural decoration to essentially engineering structures, the Tower Bridge, London, being the 'tour-de-force' of this kind of treatment. Stephenson and Telford had gothic portals added to their railway and road bridges at Conway. Telford added similar gothic turrets at each end of his splendid Craigellachie arched iron bridge. More recently, the long-span steel-arched bridge between Newcastle and Gateshead was provided at each end with pairs of massive neo-classical pavilions.

Fig. 13.22 North Road Bridge, Exeter

Fig. 13.23   Remnant of London Chatham and Dover Railway bridge, London

EMBLEMS

Emblems on our older bridges are relatively rare, but need to be preserved where they still exist. One of the earliest set is on the Brig O'Dee at Aberdeen, where inscriptions and coats-of-arms of Bishop Dumbar (1520–1627) are cut in stone on the cutwaters.

The Cambridge college bridges have coats-of-arms and emblems relating to each college, for example the portcullis carved on the parapet and gate piers of St John's Bridge. The keystone of bridge arches was a favoured spot for emblems, for instance the garlanded heads on Magdalen Bridge, Oxford.

With the introduction of cast iron, emblems became more prolific on bridges. Many of the Thames bridges bear coats-of-arms of the cities of London and Westminster and the London boroughs. Railway bridges also have their emblems, one of the most elaborately spectacular being of the (former) London, Chatham and Dover Railway on the abutment of a former railway bridge next to Blackfriar's Bridge, London, Fig. 13.23, although both the bridge itself and the railway company have long since disappeared! It is encouraging that the large cast-iron structure, dated 1864, is still kept well painted.

The national emblems, rose, thistle, leek and shamrock in cast-iron on Betws-y-Coed Bridge have already been mentioned. The unusual coat-of-arms on Homersfield Bridge is included in the section on bridge ancillaries. Like the Thames bridges, many bridges up and down the country bear cast-iron plates of the boroughs in which the bridges are located.

Canal bridges are all identified by cast-iron number plates, some also bearing the legends of the original canal companies.

PAINTS AND PAINTING

Iron and steel bridges depend for their longevity to a great extent upon a regular regime of anti-corrosion painting. Indeed, this is usually the largest single item of maintenance cost. Methods of safely minimising this cost, therefore, without significantly lowering the degree of protection afforded by paint, are worth careful analysis.

Historic bridges containing metalwork were often not conceived in this way, although there are some notable exceptions. 'Painting the Forth Bridge' is a well-known and acknowledged task that has been continuously carried out since the bridge was completed in 1890.

Metal parts of all historic bridges should be assessed in terms of:

- Their 'natural' rate of corrosion, noting that mild steel will oxidise more quickly than cast iron and that wrought iron has a very variable rate of corrosion dependent upon the quality of its initial forging.
- Conditions that may increase or reduce the natural rate of corrosion. Water-holding pockets will tend to increase the rate. Quick-drying surfaces and surfaces from which air and water are excluded will be less susceptible to corrosion.
- Their accessibility and the frequency of access currently required to maintain that part in good order.
- Current availability of access equipment, either 'built-in' as in safety harness cleats or free-standing as in hydraulic-lift platforms.
- Identification of critical points in the structure, where excessive corrosion could be catastrophic.

A regime of paint protection can then be established, providing permanent access facilities where this is economically justified. Critical areas, once identified, should not be neglected, but many cast-iron surfaces can be left for long periods if they are non-critical or have been well protected in the first place. In the nineteenth century, it was common practice to dip cast iron members in boiling linseed oil. This permeated the pores of the iron and in sheltered locations still seems to offer excellent protection.

Traditionally, iron and steel were protected, after removal of mill scale, by a heavy coat of lead oxide primer ('red-lead'). Thereafter several undercoats of oil paint were applied to protect the red-lead, followed by one or more coat of weather-resistant impermeable gloss paint. At the present, chlorinated rubber paint is found to give good protection, but this, like lead oxide, is soon to be withdrawn from general use on environmental safety grounds. Lead paints are still, currently, allowed by licence for use on historic buildings and it is hoped that the continued use of adequately protective paints will be permitted for historic bridges.

Metal bridges offer considerable scope for colour inventiveness, not always taken. Utilitarian girder bridges are often painted all over grey or black, but this need not be so. For many years now, the London metal bridges, including Holborn Viaduct and two suspension bridges, Albert and Chelsea, have been variously decorated in many different colours. Hungerford Railway Bridge was, until recently, decorated in colours that identified each structural element, an attractive idea that helped observers to understand the structure.

Other bridges have their architectural detail highlighted as in interior decoration. This can be overdone. One needs to recognise, as in plasterwork, that light (sunlight or artificial), falling on the differently inclined surfaces of a moulding will give different tonal strengths to the same colour, enhancing the three-dimensional effect of the moulding. This subtlety can be negated if the mouldings themselves are picked out in too many different colours.

Westminster Bridge is a good example where the florid multitone painting has been replaced by the original single colour of paint. The shadow effect of the modelling gives the required differences in tone to emphasise the architectural modelling. Similarly the overdone gilding has been edited back to the original intention, where it makes the details of the carved roses and lantern ornament sparkle, thus emphasizing the form of the bridge, making it look elegant rather than brassy.

Bridges with many ledges can attract the unwelcome attention of birds who perch or nest, unless deterred, and create an unhealthy build-up of guano, Fig. 13.24. This can be unhealthy for the structure, because of the quantity of damp and corrosive material, and unhealthy for maintenance operators because of the risk of disease (the writer was laid up for 6 months after catching psittacosis, now known as ornithosis, from pigeon guano). Any such accumulation will require specialist removal to avoid a health hazard.

Fig. 13.24 Birds can create an unhealthy build-up of guano

Deterrent measures should then be installed, again by specialist contractors, that are effective in the long term and which do not inhibit access for regular maintenance including painting. Spikes tend to become filled with debris, then are ineffective. Black nylon netting is surprisingly inconspicuous and if securely fixed, yet made removable, can offer a good solution to the problem.

Alternatives to painting are sometimes appropriate, especially to the stranded cables of suspension bridges where corrosion can occur in places inaccessible to paint. Some bridges have cables wrapped in several layers of protective tape which excludes water better than paint, but which still requires regular maintenance.

BRIDGES AND BRICK DECORATION

Some of the earliest Roman bridges and aqueducts are built in brick, but brick-making fell out of favour until the late medieval period and even then, stone masonry was the normal bridge building material until the late eighteenth century.

In the eighteenth and early nineteenth century, bricks were first used as 'corework' in bridges and buildings generally, faced with thin layers of stone not infrequently held together by iron cramps. Bridges, apparently built in stonework but where stones are damaged by cramp corrosion, may be of this type of construction and less decay can be tolerated than in solid masonry bridges.

Bricks used in corework were often under-fired 'seconds' or rejects and their deterioration caused by damp penetration may also be a cause for concern.

First of all in domestic buildings, then in public edifices and eventually in bridges, brick came to be accepted as an appropriate facing material, although it was usually regarded as slightly inferior to stone. The unassuming, pleasantly designed bridge at Great Wishford, Wiltshire, Fig. 13.25, is a good example. Built generally in brick, and with attractively rounded piers, stone is nevertheless employed for copings and other dressings.

Brick really came into its own in the great eras of canal, then railway construction. The versatility of 'standard' brick units was recognised. The durability of bricks improved with the introduction of 'engineering' or hard-fired blue bricks.

Brick arches and barrel vaults could easily be turned on centring using 'standard' bricks, rather than specially cut stones. This was especially useful in the case of 'skew' bridges, where the river, road or railway crossed is not at a right angle. The projecting tails of skewed brick courses often provided a decorative facing detail to an otherwise plain bridge.

At Calstock viaduct, the principal barrel vaulted arches are turned in stone, but one arch is skewed over a pre-existing roadway and is turned in brickwork, Fig. 13.26.

Because of the many different types of brick employed in bridge construction, in carrying out repairs, it is very important that replacement bricks should be of closely matching colour, texture and general durability to the original material. As noted for stonework, use of the correct type of mortar and finish of the pointing is always critical. Mortar should always be less

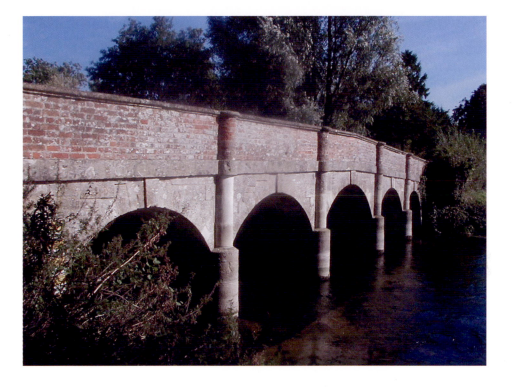

Fig. 13.25 Great
Wishford bridge,
Wiltshire

Fig. 13.26 Calstock
viaduct

hard, when set, than the material it separates and should not project beyond the surface. Hard, projecting, 'ribbon' pointing, as well as being unsightly, can trap water which, on freezing, will damage the brickwork more than if there were no pointing at all!

Most canal bridges built entirely of brick, but with stone copings, have survived some 200 years really well, which is a tribute to their good initial materials and construction. Their barrel vaults may be turned in one of two ways: either by three or four courses of brickwork laid one above the other, or by a single course one or two bricks deep set at right angles to the intrados of the arch. A few canal bridges were built to a more elaborate design than the 'standard' bridge. At Cosgrove, Northamptonshire, the road bridge over the canal was built to celebrate the opening of the Grand Junction Canal in 1800. It is faced in stone with an enriched four-centred arch flanked by turrets and both 'blind' and 'recessed' niches. The Regent's Park brick canal bridge designed by James Morgan, is embellished with bull's eye recesses in the spandrels and massive cast-iron columns or pilasters supporting the springing of the vault. John Rennie used similar columns decoratively at the old Waterloo Bridge, London (now replaced), and at Kelso, Roxburgh (still extant).

Some of the greatest railway viaducts were also built in brick, including at Welwyn, Hertfordshire and across the Welland Valley, Northamptonshire. Whilst these and other railway bridges still in use continue to receive regular maintenance, some no longer used by the railway companies, because the rail track has been lifted and the land sold off, will need continuing care, even in alternative use as footpaths or cycle ways. Drainage channels need to be kept clear of vegetation or water can build up behind spandrel walls and abutments, which on freezing, can badly damage facing brickwork.

Smaller nineteenth and twentieth century bridges, whether of all masonry construction or of composite construction with iron or steel girders or trusses, sometimes have decorative brick piers, for instance diaper work of different coloured bricks, or by the use of recessed or projecting panels. Care should be taken during repairs to ensure that such patterns are maintained. Lutyen's use of brick in bridges has already been mentioned. He often used thin bricks with relatively wide joints, quite unlike the canal and railway bridges, which were normally built in large bricks with narrow joints. Lutyen's brickwork quickly took on a 'rustic' flavour which is quite difficult to match when repairs are necessary. It is vital that replacement bricks should be handmade to the same gauge as the originals; it is not satisfactory to cut down 'full' bricks because of the harsh edges that would be produced.

AESTHETICS OF CONCRETE

With the exception of a few notable examples, concrete bridges, have generally failed to achieve the aesthetic appeal of steel and masonry. This is probably due to a variety of reasons. Until recent years the more scenic location and longer crossings that showed bridges to best advantage have been bridged by masonry or steel, leaving concrete to be used for the shorter spans in less attractive locations. The aesthetics of concrete bridges in the

Fig. 13.27 Montrose Bridge, Scotland

mid-twentieth century tended to be more concerned with simplicity and elegance, e.g. Maillart's bridges and Kingsgate Bridge, Durham.

Cracks, leakages from expansion joints, etc., and rust staining are other factors that detract from the appearance of concrete. Furthermore, concrete lacks appealing features such as the irregularities and mortar joints of masonry or the intricacies and different painted finishes of iron and steel construction. In consequence, the layman rarely takes much interest or becomes fond of concrete bridges. In fact, most people easily recall and recognise Thames bridges, such as Tower Bridge and Westminster Bridge, but few could as easily recall even such a fine structure as the present London Bridge. The explanation may in part be due to the economic and minimalist approach to design in the last 60 years which provides little that can be related to on a human scale.

However, attitudes to concrete bridges and their maintenance have changed in recent years and it is now being recognised that there are concrete bridges having both historic value and intrinsic appeal as exemplified by the grade II heritage listing of bridges such as Gifford's Rhinefield Bridge, Hampshire (Fig. 4.15) and Phillips' Homersfield Bridge, Norfolk (Fig. 4.12). Sir Owen Williams designed several fine bridges in the 1930s, including Montrose Bridge, Scotland (1930) to replace a nineteenth century wrought iron suspension bridge. The appearance of his reinforced concrete bridge reflects the profile of the suspension bridge, see Fig. 13.27.

BIBLIOGRAPHY

'The Aesthetics of Concrete Bridges', 2000 Concrete Bridge Development Group Technical Guide 4 (ISBN 0 946691 80 0), Crowthorne.

Chapter 14

Archaeology

Archaeological recording and interpretation of complex structures, such as a bridge, should be viewed as an essential part of any programme of conservation. Such recording and interpretation can assist in achieving a better understanding of the origin and development of structures through time, which can often be of significant use in designing new works or repairs.

RECORDING ARCHAEOLOGY

At the simplest level, such recording is likely to entail historic and cartographic research in such sources as county and national heritage records, record offices, and journals and periodicals. Original drawings (architects and engineers) and specifications may sometimes be located, but more often it is possible to locate and use contemporary accounts of the construction. Both sources may reveal information hidden from a current inspection by more recent additions or modifications. It is also nearly always necessary to produce a comprehensive photographic record and drawings of the bridge 'as found', and these should be updated and annotated throughout works to provide a complete record of the new works, as well as of the original. Both photographs and drawings, in plan and elevation, should be to suitable standard scales such as 1:20 and 1:50. The former Royal Commission on the Historic Monuments of England produced an excellent guide to the elements of a good structural record, see Bibliography for references to guides to site recording.

Non-intrusive testing, such as surface-penetrating radar, is increasingly being used to assist in the design of repair works, and the results are equally useful to the archaeologist in revealing hidden elements of structures.

Intrusive exploration of a structure is often required by the engineers to define the precise extent of repairs, and these works should be at least monitored by, if not undertaken by, archaeologists. This is essentially an excavation, and whilst the archaeological work may vary from or extend the engineers' requirements, the information arising from such work will be equally valuable to both professions.

The archaeologists should also be retained to maintain a continuing record of the historic elements of a structure throughout the works, in order to produce as complete a record as possible. At the end of the project, the archaeologists should prepare a report and archive (graphic and artefactual) and deposit it with some appropriate repository (e.g. a county record office and the national Monuments Record of English Heritage), where it will be accessible to the next generation of architects, engineers and archaeologists. The Highways Agency is now assembling an electronic database of all archaeological studies carried out for its roads and bridges.

The most influential policy guidance affecting the conduct of archaeological work is contained within *Planning Policy Guidance Note 16: Planning and Archaeology* (1990) and *Planning Policy Guidance Note 15: Planning and Historic Environment* (1994). The majority of work on bridges will be conducted outside the scope of either guidance note (as they pertain to the processes of seeking planning consents through the Town and Country Planning Acts), but the processes promulgated are nonetheless relevant and applicable to bridges. Other works associated with bridge repairs (e.g. alternative roads, construction compounds, etc.) are more likely to be controlled within the planning control process and both PPG15 and PPG16 may be applied to them. In particular, the sequence or stages of gaining progressively more and more detailed information is common to most archaeological and historical investigations. Another relevant document, with no statutory authority, is *Management of Archaeological Projects* (2nd edition, 1991) published by English Heritage. This document outlines in considerable detail the objectives, methods and performance indicators for the conduct of archaeological investigations. A familiarity with the stages of work will benefit all parties working on historic bridges, as the stages may be quite lengthy.

As an example of archaeological recording, scheduled monument consent (SMC) was obtained by Durham County Council in 1993 for extensive restoration works to Whorlton Bridge across the River Tees including replacement of the wooden decking, retensioning of the suspension elements and repainting of the nineteenth century iron works. This work was undertaken between 28 September 1993 and completed by 20 December the same year. Archaeological recording consisted of a pre-SMC application meeting between the County Archaeologist, County Bridges Engineer and the regional English Heritage Inspector of Ancient Monuments to discuss the extent of works and need for recording. The level of recording felt necessary at this point was limited to a comprehensive photographic record of the bridge before and after the scheme of restoration. This was duly undertaken and the photographic archive deposited with the County Sites and Monuments Record in January 1994.

In addition to its cultural importance, the archaeological record can be useful in an engineering sense. On occasions when a structure is seriously damaged, for example floods may displace and destroy some of the stonework of a masonry bridge, the record can be consulted to enable the stones to be replaced in their original positions.

ARCHAEOLOGY OF FOUNDATIONS

Conservation of an historic bridge may require strengthening of its foundations, but this could destroy significant buried archaeological evidence. Piers, timber piles, rafts and inverts are all significant features that can yield important information about the history of the bridge. Dendrochronological analysis (tree ring dating) of preserved timbers can provide an accurate date of early foundations. Furthermore, post-medieval and even medieval bridges may have been constructed on the foundations of earlier bridges. This reinforces the need for a preliminary desk study of the site before planning the conservation work. Wherever possible, the engineering work should be designed to preserve historic features irrespective of whether they are buried or visible. On historic crossings, it is also possible that the present bridge may have been built beside an older bridge that was replaced, or it may have been built beside a ford. In either case, care should be taken to avoid disturbance to the bed of the river or its banks, including either side of the present bridge.

When the A4130 was diverted around the historic medieval bridge in Wallingford (Oxford), the most appropriate locations for the new crossing over the River Thames coincided with a Bronze Age riverside settlement dated to circa 2000 BC. Archaeological excavations, small in extent but of considerable depth, allowed the part of the settlement that would be affected by construction works to be recorded, providing much new information about European trade and settlement patterns of the second and third millenniums BC, see Fig. 14.1.

ARCHAEOLOGY OF MASONRY ARCH FILL

The fill material within masonry arch bridges is sometimes composed of recycled material and debris that was to hand at the time of construction. Such material may contain archaeological artefacts that can provide information about the date of construction and conditions of the time. It follows that on occasions when the fill is being excavated for one reason or another, there should be an archaeologist present to observe and record information. This was done on the occasion when Woolbeding Bridge in West Sussex was strengthened by the addition of concrete saddles over the arches, see Fig. 14.2. The engineering activities required removal of all the fill and exposure of the intrados of each arch. This provided a unique opportunity to carry out a full investigation of the constituent materials in the fill. As Woolbeding Bridge is a late medieval structure built in the sixteenth century, it was possible that there could be some interesting discoveries. In the event the fill was found to be original and undisturbed. Some small artefacts were discovered, the oldest being Roman, and useful information was obtained in relation to successive running surfaces which had been laid one on top of the other at intervals of about 100 years. In one of the running surfaces, it was possible to identify ruts made by cart wheels hundreds of years ago.

Sadly, the fill in many arch bridges has been disturbed in the past, either by the digging of trenches for installation or reinstallation of utilities, or for strengthening the load carrying capacity by one means or another and with-

Fig. 14.1 Excavation in Wallingford.
Archaeological investigation of Bronze Age riverside settlement prior to construction of new crossing of the River Thames

out archaeological observation. This has destroyed the archaeology and degraded the intrinsic quality of the structure.

ARCHAEOLOGY OF ASSOCIATED STRUCTURES

Medieval bridges sometimes had associated structures such as fortifications, chapels, etc., see Chapter 5. A few survive, but most have been demolished or incorporated into more recent construction. The survivors are self-evident and well known, so that there should be little risk of failing to identify them. Where associated structures existed in the past there may be remnant foundations or other components that should be left undisturbed or conserved.

Elvet Bridge in Durham city provides an excellent example of publicised archaeology. The original bridge was built in the twelfth century, reconstructed in 1228 and much repaired and altered since. At the south-east end of the bridge, fragments of the original twelfth century bridge and the Chapel of St Andrew are visible. This history is outlined in a plaque fixed to the bridge, see Fig. 14.3.

Elvet Bridge has a history typical of medieval masonry in that it has 'lost'

Fig. 14.2   Woolbeding Bridge, West Sussex. Archaeological investigation carried out during the strengthening work

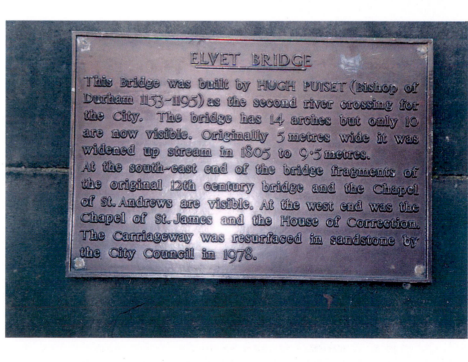

Fig. 14.3   Elvet Bridge, Durham city. Plaque fixed to the bridge to inform on its history

several of its original arches (three having been destroyed by floods in 1771) and it was widened in the nineteenth century.

Before planning remedial works, an archaeological desk study should be carried out to investigate the past history of the bridge and provide early warning of the likelihood of any archaeological remains being present. There are many examples of medieval bridges located in town centres which have been widened on one side or both so that most of the original material is hidden from view. These bridges are as worthy as others, because it is always possible that they could be uncovered in the future if, for example, the road is re-engineered or rerouted. Chantry Bridge Rotherham and Morton Bridge in Warwickshire are cases where original elevations have been retrieved in this manner.

BIBLIOGRAPHY

Royal Commission on the Historic Monuments of England. (1996) *Recording Historic Buildings: A Descriptive Specification*, 3rd edn, RCHME.

Spence, C. (1990) *Archaeological Site Manual of the Department of Urban Archaeology Museum of London*. The Museum of London.

Department of the Environment 'Planning and Archaeology' Policy Planning Guidance Note 16 HMSO 1990, and 'Planning and Historic Environment' Note 15 1994.

English Heritage. (1991) *The Management of Archaeological Projects*.

Appendix

Highways Agency bridges constructed up to 1915

Highways Agency bridges date back to the fifteenth century and some originated earlier. They encompass a variety of size and form of structures, the majority of those built before 1915 are of masonry construction.

Altogether the Highways Agency is responsible for some 8,124 bridges (National Structural Database, 2000), of which 339 were constructed before 1915. The structures are generally road-over-river or road-over-rail and vary in size from single-span to eleven-span viaducts. Various materials have been used in the construction of the bridges including stone, brick, slate, iron and steel.

The bridges are listed in Table A1, which provides an overview of the historic structures on the highway network. The bridge location, construction date and generic construction type are outlined with brief modification details. 'Visibility' provides an indication of how prominent the structure is, in relation to nearby dwellings, navigable waterways or from the structure itself. The 'Visual Condition' identifies whether the original construction has been retained or has been obscured by any modifications. Some modifications such as deck strengthening may have no impact on a bridge's visual condition.

Dates of construction are usually taken as the earliest recorded reference to the bridge, but are questionable as they could be related to an earlier structure. In cases when there are no credible records the dates have been estimated from the appearance and form of construction, hence many are broadly dated as, for example, twenty-five which are entered as 1800.

The majority of the bridges built before 1915 are masonry arches constructed in brick or stone. Originally they were rather narrow and many have had to be widened, sometimes on several occasions, in order to meet the requirements of increased volumes of traffic. These modifications are represented in Fig. A1 and summarised as follows:

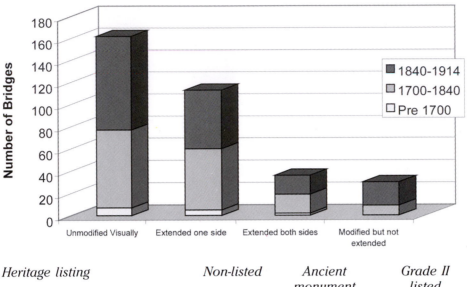

Fig. A.1 Histogram of modifications

Heritage listing	Non-listed	Ancient monument	Grade II listed
No change to the width	174	10	6
Widened on one side	106	4	2
Widened on both sides	35	—	—
Total	315	14	8

There is a total of 296 masonry arch bridges. It can be seen that fourteen of the bridges are national monuments and four of these have been widened on one side. Eight are grade II listed and three have been widened. These figures take into account that Marton Bridge, an ancient monument constructed by 1414 and probably much earlier, had a reinforced concrete widening removed in 2000.

Marton bridge was constructed by 1414 and is a masonry arch structure across the River Leam in Warwickshire. The bridge was widened in 1926 to accommodate the ever-increasing traffic demands. However, in 2000, a new structure has been built alongside, and the widening removed. The original fabric of its construction has been exposed to restore the bridge to its former beauty.

All of the 138 bridges built between 1700 and 1840 are of masonry arch construction. Wreaks Causeway is a good example of a slate masonry arch, which is a three-span structure carrying the A595 over Kirby Pool in Cumbria. This bridge is of fine appearance and is set in beautiful surroundings.

Whatstandwell Bridge in Derbyshire is a stunning seven-span masonry arch bridge, which has been classified as an ancient monument. The bridge was constructed in 1795 and carries the A6 main road. It is in good condition and in its original state, adding greatly to the character of the surrounding area.

The beginning of the nineteenth century saw the first railway bridge construction, which can be illustrated by Scarrow Hill Railway Bridge on the A69, built in 1838 in Cumbria. The single-span masonry bridge has been strengthened in 1956 by the addition of a concrete saddle, although visually it remains in its original condition.

Tempsford in Bedfordshire is the site for three ancient monuments on a short stretch of the A1. The North Flood Arches, South Flood Arches and Tempsford Bridge were all constructed around 1815 to span the navigable Great Ouse River and its associated flood plain. These impressive multispan stone arch bridges are in good condition with only minor deck strengthening to the main bridge.

Concrete bridge extensions can be effectively dressed by the use of facing materials to match the existing structure and the relocation of the original parapet, as used in the widening of Blackwell Bridge in North Yorkshire. This three-span arch structure built in 1832 across the river Tees has had a stone facing on the extended elevation to match the original and reduce the impact of the widening work.

1838 saw the construction of Limpley Stoke Viaduct in Somerset, which carries the A36 across the River Avon. The eleven-span viaduct remains unaltered with its original parapets and is a fine example of this construction technique.

Cast and wrought iron bridge deck construction incorporating transversely spanning jack arches became fashionable around the middle of the nineteenth century. Carlton Station Railway Bridge in North Yorkshire is a good example of this type of construction and was built in 1882. The bridge incorporates wrought iron beams with brick jack arches and remains in its original condition.

Pinfold Bridge is an *in situ* concrete portal frame, beam-and-slab construction. This 1913 canal bridge incorporates masonry parapets, which reduces the impact of the early concrete.

The majority of the Highways Agency structures detailed are in their original condition and many fine examples of historic value can be seen on the road network today. These bridges are of national importance and every effort is made to conserve them.

Table A.1. Highways Agency bridges constructed before 1915

Year	Structure name	Road	County	Original Bridge Form/materials	Extension One side	Extension Both sides	Modifications	Visibility	Visual condition	Importance	Heritage listing
1272	Llangua	A465	Hereford & Worcester	4-span stone arch	Y		Widened	High	Poor	V high	Grade 2
1414	Marton Bridge	A423	Warwickshire	Twin span masonry Bridge			Pedestrianised	Medium	Good	Medium–low importance civic bridge	Ancient monument
1430	Matlock	A6	Derbyshire	3-span stone arch	Y		Widened	High	Good	V high	No
1450	Hanging Bri f/arch-m/arch	A52	Staffordshire	Stone 3-span arch		Y	Widened	Medium	Good	V high	No
1450	Hanging Bri m/race-n/arch	A52	Staffordshire	Stone Single arch	Y		Widened	Medium	Good	High	No
1490	Hanging Bridge, orig	A52	Staffordshire	Stone 2-span arch		Y	Widened	Medium	Good	V high	No
1500	Monks Bridge	A38	Derbyshire	4-span masonry arch			RC saddle	High	Original	Medium important civic bridge	Ancient monument
1597	Prospect Flood arch	A40	Hereford & Worcester	Twin-span stone arch				High	Good	V high	No
1597	Cleeve	A40	Hereford & Worcester	5-span stone arch				High	Good	V high	No
1597	Wilton Flood arch	A40	Hereford & Worcester	Single-span stone arch	Y		Widened	High	Medium	High	No
1597	Wilton Bridge	A40	Hereford & Worcester	6-span brickwork and masonry arch	Y		Rolled steel section & concrete extension	Medium	Original one side, new parapet on other elevation	Medium–low importance civic bridge	Ancient monument
1660	Heathfield Rly M/arch	A38	Devon	Stone arch	Y		Widened	Medium	—	Medium	No
1700	Claymills (arch orig)	A38	Staffordshire	Single-span brick/stone arch	Y		Widened	High	Good	High	No
1700	Beeston A	A49	Cheshire	Stone				High	Good	High	No
1730	Scampston	A64	North Yorkshire	3-span classical stone Structure				High	Good	V high	No
1750	Spithopehaugh	A68	Northumberland	Stone packhorse bridge		Y	Widened	Hidden	Poor	Low	No
1750	Priests Bridge	A1	Northumberland	Stone arch	Y		Widened	Dramatic stone wall	Slight spalling	Medium	No
1751	South Great Glen Old	A6	Leicestershire	Brick arch	Y		Widened	Low	Good	Low	No
1751	Great Glen Bridge	A6	Leicestershire	5-span segmental brick arch			Pipeline added	Medium	Original	Medium importance civic bridge	No
1756	Longtown Bridge	A7	Cumbria	5-span masonry arch	Y		Widened 1889	High-from footpath	Original one side	Medium–high	Grade 2

Date	Name	Road	County	Construction	Altered	Alteration	Significance	Condition	Importance	Listed
1770	Stambridge	A4	Avon	Stone culvert			High	Good	Medium	No
1770	Lambridge	A4	Somerset	Stone arch			Medium	Good	Medium	No
1770	Newbridge	A4	Avon	Stone arch			High	Good	V high	No
1775	Blue	A19	N. Yorkshire	Stone			Yes, Rural	Good	Medium	No
1775	Golbourne Bridge	A41	Cheshire	Masonry arch		Concrete in-fill and backing for highway	Obscured by vegetation	Original	Medium–low importance civic bridge	No
1775	Spital Kilvington Bridge	A19	N. Yorkshire	Masonry arch	Y	In situ R.C. solid slab extensions & new deck	Low	Original structure hidden both sides	Minor culvert	No
1777	Wreaks Causeway Bridge	A595	Cumbria	3-span Slate masonry arch		Concrete saddle added	Medium	Original	Medium importance civic bridge	No
1780	Filleybrooks River	A34	Staffordshire	Brick	Y	Widened	Good	Good	High	No
1780	New Trent	A34	Staffordshire	Brick/stone culvert	Y	Widened	Poor	Good	Low	No
1780	Pickford	A45	West Midlands	Twin-span brick/fab. steel culvert	Y	Widened	Medium	Good	Low	No
1780	Brailsford	A52	Derbyshire	Stone arch			High	Good	High	No
1780	Mill House-Adlington	A523	Cheshire	Stone arch			High	Good	High	No
1780	Meols	A565	Lancashire	Masonry arch	Y	Widened	High	Good	High	No
1780	Robbins New (m/arch)	A59	West Yorkshire	Stone arch	Y	Widened	Low	Poor	Low	No
1780	Tarleton Canal	A59	Lancashire	Stone arch			High	Good	Medium	No
1780	Double arch	A59	Lancashire	Stone arch			V high	Good	V high	Grade 2
1780	Crickle	A59	Lancashire	Stone arch			High	Good	V high	No
1780	Duffield Town	A6	Derbyshire	Brick/masonry/stone	Y	Widened	High	Good	V high	No
1780	Crickle Bridge	A59	N. Yorkshire	Masonry arch		2 retaining walls added to west	Medium	Original	Minor culvert	Ancient monument
1780	Thorlby Sheep Creep	A65	North Yorkshire	Stone			Medium	Good	Medium	No
1780	Marsh Beck	A660	West Yorkshire	Stone arch			High	Good	High	No
1780	Desborough South Bridge	A6	Northamptonshire	Twin-span masonry arch	Y	Mass concrete extensions either side	High	Original hidden	Medium importance civic bridge	No
1785	Hoppersford (old)	A43	Northamptonshire	Brick arch			Medium	Original	Minor culvert	No
1789	Dunton Wharf Bridge	A446	Warwickshire	Brick arch	Y	Extended twice, in situ R.C. simply supported solid slab extensions	High from canal & pathways	Original one side	Medium importance civic bridge	No
1789	West Bridge	A614	Nottinghamshire	3-span masonry arch		Concrete backing & inverts added	High	Original	Medium importance monument	Ancient monument

Table A.1. Continued

Year	Structure name	Road	County	Original Bridge Form/materials	Extension One side	Extension Both sides	Modifications	Visibility	Visual condition	Importance	Heritage listing
1790	Crambeck Bridge	A64	N. Yorkshire	6-span masonry arch	Y		Steel plate girder and troughing R.C. saddle	Medium	Original one side	Medium	No
1790	Wadesmill	A10	Hertfordshire	6-span brick/stone arch				Obscured by vegetation	Good	V high	No
1790	Nell Lock Canal Bridge	A41	Oxfordshire	Masonry arch	Y		Prestressed beam extension	High from canal	Original one side	Medium–low importance civic bridge	No
1793	Strongford Bridge	A34	Staffordshire	Masonry arch		Y	R.C. simply supported extension on both sides	Medium	Original hidden by new simply supported R.C. extension	Major civic Bridge	No
1795	Whatstandwell Bridge	A6	Derbyshire	7-span masonry arch			Retaining wall added next to bridge	Medium-high	Original	Medium importance civic bridge	Ancient monument
1796	Healam Old	A1	Yorkshire	Masonry arch		Y	Widened in similar style to original	Medium	Extensions each side in style of original	Minor culvert	No
1796	Ryton River Bridge	A45	Warwickshire	3-span brick/masonry arch	Y		Extended twice on one side, *in situ* R.C.	Medium	Original, one side	Medium–low importance civic bridge	No
1797	Weeford Northbound Orig	A38	Staffordshire	Stone arch				High	Good	High	No
1797	Wychnor Bridge	A38	Staffordshire	3-span masonry arch	Y		*In situ* R.C. extension	Medium	Original one side, concrete other	Medium importance civic bridge	No
1800	Ease Drain	A1	Lincolnshire	Brick	Y		Widened	Rural	3 arch, good	Minor	No
1800	Humber Head Dyke-M/arch	A1	W. Yorkshire	Stone		Y	Widened	Hidden	Poor, lacking	No	No
1800	The Tunnel	A483	Shropshire	Twin-span stone culvert				Medium	Good	Low	No
1800	Gailey Canal (old)	A5	Staffordshire	Brick	Y		Widened	V low	Poor	Low	No
1800	Accommodation	A58	W. Yorkshire	Stone arch				Low	Medium	Low	No
1800	Dam Brook	A59	Lancashire	Stone arch				Low	Good	Medium	No
1800	Roam (orig m/arch)	A590	North Yorkshire	Stone arch		Y	Widened	Hidden	—	Low	No
1800	Warney	A6	Derbyshire	Stone				Low	Poor	Medium	No
1800	Cliviger	A646	Lancashire	Stone arch				Hidden	Poor	Low	No
1800	Redwater Clough (M/arch)	A646	West Yorkshire	Stone culvert		Y	Widened	Hidden	Poor	Low	No

Year	Name	Road	County	Type		Modification	Visibility	Condition	Importance	Listed
1800	Scaitcliffe (m/arch)	A646	West Yorkshire	Stone culvert	Y	Widened	Hidden	Poor	Low	No
1800	Hawks Clough Goit arch	A646	West Yorkshire	3-span stone arch			High	Poor	High	No
1800	Luddenden Foot (m/arch)	A646	West Yorkshire	Stone culvert	Y	Widened	Low	Poor	Low	No
1800	Ottercops (m/arch)	A696	Northumberland	Stone arch			Medium	Good	Low	No
1800	Birdholme Bridge	A61	Derbyshire	Masonry arch	Y	R.C. simply supported extension	Low	Original	Low importance civic bridge	No
1800	Bridge Hotel	A48	Gloucestershire	Masonry arch	Y	Solid slab R.C extension	Medium	Original one side	Medium	No
1800	Burton Overy Parish Boundary Bridge	A6	Leicestershire	Twin-span masonry arch			Medium, High	Original	Medium importance civic bridge	No
1800	Hambrook House	A40	Gloucestershire	Brick and masonry arch			Original hidden by vegetation	Original	Minor culvert	No
1800	Hesketh Bridge	A5	Northamptonshire	Brick/masonry/stone arch			Low, hidden by vegetation	Original	Minor culvert	No
1800	Mytholmroyd Canal Bridge	A646	W. Yorkshire	Masonry arch	Y	concrete encased steel beam footpath widening	High	Original one side	Medium–low importance civic bridge	No
1800	Sandford Mill Road Bridge	A40	Gloucestershire	Brick arch			Medium–low, cannot be seen from main road	Original	Minor culvert	No
1800	Towcester Bridge	A5	Northamptonshire	Twin-span Brick arch			Low, hidden by vegetation	Original	Minor culvert	No
1800	Turvey Bridge	A428	Bedfordshire	11-span masonry arch	Y	R.C. solid slab extension	Medium	Original one side	Medium importance civic bridge	Ancient monument
1800	East Morton	A59	N. Yorkshire	Double-Deck masonry arch			High		High	Grade 2
1800	Weedon Canal	A45	Northamptonshire	Masonry arch	Y	W.I./Jack arch extension	Low	Original one side	Medium	No
1801	Onibury Mill	A49	Shropshire	Random rubble masonry arch			Low	Original	Minor culvert	Unknown
1807	Bargate Bridge	A16	Lincolnshire	Masonry arch	Y	R.C. arch extension	High	Original one side	Major civic bridge	Grade 2

Table A.1. Continued

Year	Structure name	Road	County	Original Bridge Form/materials	Extension One side	Extension Both sides	Modifications	Visibility	Visual condition	Importance	Heritage listing
1809	Bodley Brook (M/arch)	A30	Devon	Single-span brick/stone arch	Y		Widened	Rural low	Good	No	No
1809	Fenny Bridge	A30	Devon	3-span brick arch				Medium	Original	Medium importance civic bridge	No
1810	Wadesmill Hill	A10	Hertfordshire	Brick arch				Rural	Good	Minor	No
1810	Radford Canal Original	A34	Staffordshire	Brick arch	Y		Widened	Poor	Poor	No	No
1810	Rushton Canal Feeder	A523	Staffordshire	Stone arch				Low	Good	Medium	No
1810	Hugg	A523	Cheshire	Stone arch				High	Good	Medium	Grade 2
1810	Bosley or Linfoot	A523	Cheshire	Stone arch				Medium	Good	Medium	No
1810	Radford River Bridge	A34	Staffordshire	3-span brick arch			R.C. footpath added adjacent to structure	Medium, obscured on one side by footbridge	Steel parapets added	Medium importance civic bridge	No
1811	Wayford Bridge	A49	Shropshire	Masonry arch	Y		R.C. arch widening 1929	Medium	Original one side	Medium–low importance civic bridge	Unknown
1814	Holme Bridge	A65	N. Yorkshire	Twin-span masonry arch				Medium	Original	Medium importance civic bridge	No
1815	Bedford Town Bridge	A6	Bedfordshire	5-span masonry arch	Y		R.C. extension	High	Original	Medium–high	Ancient monument
1815	Tempsford North Flood arch	A1	Bedfordshire	7-span stone arch				High	Original	Major civic bridge	Ancient monument
1815	Tempsford South Flood arches	A1	Bedfordshire	7-span stone arch				High	Original	Major civic bridge	Ancient monument
1816	Fairfield Road masonry	A6	Derbyshire	Brick/stone arch	Y		Widened	Low	Medium	Medium	No
1818	Waterhouses	A523	Staffordshire	Single-span masonry arch	Y		R.C. solid slab arched deck	Medium	Original one side, masonry spandrel and original relocated parapet on other	Medium importance civic bridge	No
1820	Ryshworth	A650	West Yorkshire	Stone arch				Medium	Poor	Medium	No
1820	Cottingley	A650	West Yorkshire	5-span stone river bridge	Y		Widened	High	Good	V high	No

Date	Name	Road	County	Structure	Listed	Alterations	Visibility	Condition	Importance	Ancient monument
1820	Cottingley Beck	A650	West Yorkshire	Stone culvert			High	Good	High	No
1820	Rochester	A68	Northumberland	Stone arch			Medium	Medium	Medium	No
1820	Tempsford Bridge	A1	Bedfordshire	3-span masonry arch	Y	Widened R.C. Deck	High, prominent from navigable river	Original	Major civic bridge	Ancient monument
1821	Clyst Honiton Bridge	A30	Devon	3-span masonry arch		In situ concrete deck added	Medium	Original with concrete deck	Medium importance civic bridge	No
1821	Tarleton Bank Hall Bridge	A59	Lancashire	3-span masonry arch		Concrete reinforcement	Medium	Original	Medium importance civic bridge	No
1822	Chideock Road Bridge	A35	Dorset	Stone arch	Y	In situ R.C. slab extension	Low	Original one side	Minor culvert	No
1822	Eden Bridge	A66	Cumbria	4-span sandstone masonry arch			High	Original	Major civic bridge	No
1823	Newton Bridge	A4	Avon	Masonry arch	Y	Concrete arch extension & concrete saddle added	Low, hidden by vegetation	Original one side	Minor culvert	No
1823	Westlington Bridge	A7	Cumbria	Twin-span masonry arch			Medium	Original	Medium–low importance civic bridge	No
1824	Gelt (New) Bridge	A69	Cumbria	Sandstone masonry arch	Y	Widened 1824	Medium, low	Original extended	Medium importance civic bridge	No
1826	High Lane Canal Bridge	A6	Greater Manchester	Masonry arch			High, from canal & footpath	Original	Medium importance civic bridge	No
1826	Macclesfield Canal	A523	Cheshire	Masonry arch			Medium, from canal	Original	Medium importance civic bridge	No
1827	Cuttle Mill Bridge	A5	Northamptonshire	Twin-span masonry arch	Y	In situ R.C. solid slab box extension	Low, hidden by vegetation	Original one side	Minor culvert	No
1828	Grove Bridge	A49	Shropshire	Dressed sandstone arch		Footbridge added one side, 1961 (steel & R.C.)	Low	Original	Medium importance civic bridge	No

Table A.1. Continued

Year	Structure name	Road	County	Original Bridge Form/materials	Extension One side	Extension Both sides	Modifications	Visibility	Visual condition	Importance	Heritage listing
1828	Moss Pit Railway	A449	Staffordshire	Masonry arch		Y	Precast and solid slab replacement of old structure (1962)	Medium	None, original arch removed after new construction	Medium importance civic bridge	No
1829	Chappels (Lanthwaite) M/A	A595	Cumbria	Stone arch	Y		Widened	Low	Good	Medium	No
1829	Millholme (m/arch)	A595	Cumbria	Stone arch	Y		Widened	Low	Medium	Medium	No
1829	Knightsmill Bridge	A39	Cornwall	3-span masonry arch	Y		R.C. slab extension with new parapet to match existing	Low	Original one side	Minor culvert	No
1829	Muncaster Bridge	A595	Cumbria	Local Granite masonry arch				Medium	Original	Medium–low importance civic bridge	No
1830	Calder (M/arch)	A595	Cumbria	Stone arch	Y		Widened	High	Good	High	No
1830	Boot (M/arch R/C Saddle)	A595	Cumbria	Stone arch		Y	Widened	Hidden	—	No	No
1830	Sandraw (M/arch)	A596	Cumbria	Stone arch	Y		Widened	Low	Good	Low	No
1830	Tinkersley (M/arch)	A6	Derbyshire	Stone arch				Low	Poor	Low	No
1830	Cross Gates Railway No. 23	A6120	West Yorkshire	Masonry arch	Y		Masonry/steel plate girders encased in concrete	Low	Original one side	Medium importance civic bridge	No
1830	Sandford Bridge	A41	Shropshire	Masonry arch		Y	R.C. & simply supported steel & timber extensions	Medium	Original bridge surrounded by extension & footbridge	Medium–low importance civic bridge	No
1831	Lickle (M/arch)	A595	Cumbria	Stone	Y		Widened	Low	Poor	Low	No
1831	Duddon	A595	Cumbria	Stone				High	Good	V high	No
1831	Low Lickle	A595	Cumbria	Slate arch				Medium	Original	Medium/low	No
1831	Newstead Bridge	A16	Lincolnshire	Stone voussoirs & brick arch & concrete saddle			R.C. saddle added	Low, hidden by vegetation	Original	Medium	No
1832	Blackwell masonry arch	A66	N. Yorkshire	3-span masonry arch	Y		Stone-faced in situ R.C. widening of arch	Medium–High	Original one side	Major civic bridge	No
1832	Blythe West	A696	Northumberland	Masonry arch				Low-from fields	Original	Medium	No

Year	Name	Road	County	Type	Listed	Modifications	Condition	Structure	Importance	Other
1833	Crowlas Bridge	A30	Cornwall	Masonry arch	Y	R.C. extensions either side	Medium	Original structure hidden	Minor culvert	No
1833	Howend High Bridge	A7	Cumbria	Masonry arch	Y	Simply supported R.C. solid slab extension	Medium	Original one side	Minor culvert	No
1834	Shawford	A36	Devon	3-span Stone arch			High	Good	V high	No
1834	Monkton Combe (Railway)	A36	Avon	Brick arch			Hlgh	Good	High	No
1834	Monkton Combe (Canal)	A36	Avon	Stone arch			High	Good	V high	No
1834	Claverton	A36	Avon	Stone arch			Medium	Good	Medium	No
1834	Hinton Abbey underpass	A36	Avon	Masonry arch	Y	Brick and *in situ* concrete extension	High	Brick & I.C addition one side	Minor bridge	No
1834	Limpley Stoke Viaduct	A36	Avon	11-span brick/masonry/stone arch			High	Original	Major civic bridge	No
1834	River Bewl Bridge	A21	Kent	Masonry arch			Low due to surrounding vegetation	Original	Minor culvert	No
1835	Cattle Creep Bridge	A69	Cumbria	Masonry arch			Low	Original	Medium–low importance civic bridge	No
1835	Warwick Bridge	A69	Cumbria	3-span masonry bridge		Rebuilt version 1835	Medium	Original extended	Medium–low importance civic bridge	No
1836	Curdworth River	A446	Warwickshire	5-span solid slab brick arch			Medium	Original	Medium importance civic bridge	No
1837	Munton (Black Beck) m/arch	A595	Cumbria	Stone arch	Y	Widened	Medium	Good	Medium	No
1837	Glyn Bridge	A38	Cornwall	3-span masonry arch			Medium	Original	Medium importance civic bridge	No
1837	Holmrook Bridge	A595	Cumbria	Masonry arch	Y	Cased UB & R.C. slab widening	Medium	Original one side	Medium importance civic bridge	No
1838	Kirkby Thore Bridge	A66	Cumbria	Masonry arch	Y	Widened & strengthened, simply supported R.C.	Low	Original one side, other simply supported concrete deck	Minor culvert	No

Table A.1. Continued

Year	Structure name	Road	County	Original Bridge Form/materials	Extension One side	Both sides	Modifications	Visibility	Visual condition	Importance	Heritage listing
1838	Pownaughan Bridge	A69	Cumbria	Masonry arch		Y	Simply supported R.C. solid slab extension on both sides	Low	Extension both sides	Minor culvert	No
1838	Scarrow Hill Railway Bridge	A69	Cumbria	Masonry arch			Strengthened 1956	Medium–high from railway	Original	Medium importance rail bridge	No
1839	Tarvin Pool Bridge	A54	Cheshire	Masonry arch	Y		simply supported precast concrete box added	Low	Original one side	Minor culvert	No
1840	Box Brook	A4	Wiltshire	Twin-span Brick arch				Low	Poor, pipe on facade	Medium	No
1840	Goit Stock Farm Mickle Ing (m/arch)	A660	West Yorkshire	Stone culvert	Y		Widened	Low	Poor	Low	No
		A660	West Yorkshire	Stone culvert		Y	Widened	Low	Poor	No	No
1840	Bradnop	A523	Staffordshire	Stone and brick arch	Y		R.C. simply supported solid slab extension (1931)	Low	Original one side, R.C. and brickwork face on other	Minor culvert	No
1840	Bridge End	A523	Staffordshire	Stone arch		Y	R.C. arch rib and solid slab extension either side	Low	Original hidden by concrete extensions	Medium importance civic bridge	No
1840	Calderside Bridge	A646	W. Yorkshire	Masonry arch				Medium	Original	Medium importance civic bridge	No
1840	Charlestown Lower Bridge	A646	W. Yorkshire	Masonry arch				Low, obscured by trees	Original	Medium–low importance civic bridge	No
1840	Charlestown Upper Bridge	A646	W. Yorkshire	Masonry arch			Footbridge added to north side	Medium, partially hidden by vegetation	Original	Medium–low importance civic bridge	No
1840	Churchbridge Brook	A5	Staffordshire	Masonry arch	Y		R.C. solid slab extension	Low	Original north elevation, steel parapet on south	Minor culvert	No
1840	Crane Brook No.2 Bridge	A5	Staffordshire	Brick arch	Y		R.C. saddling & R.C. solid slab extension 1840	Medium	Original one side	Minor culvert	No

Year	Name	Road	County	Structure type	Widened	Widening / modification	Strengthened	Strengthening / deck work	Visibility	Original fabric	Importance	Ancient monument
1840	Milford Bridge	A6	Derbyshire	2-span masonry arch & concrete footway					High, from road	Original	Medium–low importance civic bridge	No
1840	Millstream Bridge	A596	Cumbria	Masonry arch					Low	Original	Minor culvert	No
1840	Rosewastis	A39	Cornwall	Masonry arch	Y	In situ reinforced concrete extension with masonry face			Low	Original one side	Minor culvert	No
1840	Wash Brook Bridge	A5	Staffordshire	Masonry arch	Y	R.C. solid slab box culvert extension			Low, obscured by vegetation	Original one side	Minor culvert	No
1840	Windy Railway Bridge	A646	Lancashire	Skew masonry arch					High	Original	Medium importance civic bridge	No
1840	Workington Bridge	A596	Cumbria	3-span masonry arch					Medium	Original	Medium importance civic bridge	No
1842	Jack O Watton Railway	A446	Warwickshire	Bick arch					Low, from railway	Original	Medium importance civic bridge	No
1844	Muncaster Mill Bridge	A595	Cumbria	Masonry arch					Medium	Original	Medium–low importance civic bridge	No
1845	Hope Green Railway	A523	Cheshire	Brick arch			Y	Precast, prestressed R.C. solid slab deck	Low-from railway	Original deck removed and replaced	Medium importance civic bridge	No
1845	Lady Side	M1	Northamptonshire	Masonry arch			Y	In situ R.C. solid slab extension	Low	Original mostly hidden by extension and overlay either side.	Major civic bridge	No
1846	Aspatria Railway No.49	A596	Cumbria	Masonry arch		R.C. Saddle added			Medium, highly visible from railway	Original with R.C. Saddle	Medium importance rail bridge	No
1846	Bait Hill Bridge	A11	Norfolk	Simply supported riveted plate girder with brick jack arches	Y	Cased UB with concrete jack arches and simply supported deck			Medium	Original one side	Medium importance civic bridge	No

Table A.1. Continued

Year	Structure name	Road	County	Original Bridge Form/materials	Extension One side	Both sides	Modifications	Visibility	Visual condition	Importance	Heritage listing
1846	Heathfield Rly No. 58	A596	Cumbria	Masonry arch				Medium, highly visible from railway	Original	Medium importance rail bridge	No
1847	Crossflats Railway No. 64	A650	West Yorkshire	3-span masonry/brick arches				Medium	Original	Medium importance civic bridge	Unknown
1848	Chester Road Bridge	A449	Hereford & Worcester	Brick/masonry arch			Concrete on tubular steel frame f/bridge added	Medium	Original plus footbridge	Medium importance civic bridge	No
1848	Dunston Railway	A140	Norfolk	Brick arch				Medium, highly visible from railway	Original	Medium importance rail bridge	No
1848	Nell Railway Bridge	A41	Oxfordshire	Brick arches	Y		Re-decked & widened 1958 with concrete	High	Original one side	Medium importance rail bridge	No
1849	Town Bridge (Stamford)	A43	Lincolnshire	3-span Stone arch				High	Good	V important	Grade 2
1849	Town Bridge Stamford	A43	Lincolnshire	3-span masonry arch				High, prominent feature	Original	Major civic bridge	No
1850	Chipping	A10	Hertfordshire	Brick		Y	Widened	Minor village	Poor, mutilated	Low	No
1850	Mile House	A19	N. Yorkshire	Brick arched culvert	Y		Widened	Hidden	Hidden	No	No
1850	Skew Railway	A36	Wiltshire	Single-span Brick arch				Quite High	Good	Minor	No
1850	New drain N/B, orig	A38	Staffordshire	Brick arch	Y		Widened	Low	—	Low	No
1850	Coley Beck culvert	A58	West Yorkshire	Brick arch				Low	Medium	Low	No
1850	High Peak Junction Rly	A6	Derbyshire	Stone arch corrugated Steel Structure (CSBS)	Y		Widened	High	Poor	Medium	No
1850	Lady Pitt Flood arches	A638	South Yorkshire	8-span Stone Structure				Low	Poor	High	No
1850	Lobb Mill	A646	West Yorkshire	Stone culvert				Medium	Medium	High	No
1850	Raylees East	A696	Northumberland	Stone arch				High	Good	Medium	No
1850	Kirkwhelpington (w-m/arch)	A696	Northumberland	Stone arch	Y		Widened	High	Good	High	No
1850	Allens Bridge	A556	Cheshire	Masonry arch	Y		In situ R.C. beam and slab extension added (1930)	Low	Original one side	Medium importance civic bridge	No

Year	Name	Road	County	Construction		Modifications	Visibility	Original/Extension	Importance	Listed
1850	Ambergate Bridge	A6	Derbyshire	Masonry arch	Y	PSC beams & R.C. widening	Medium	Original one side alterations hidden	Medium importance civic bridge	No
1850	Blakeney Bridge	A48	Gloucestershire	Masonry arch	Y	Extended three times	Original can be seen one side, fairly well hidden	Original one side	Minor culvert	No
1850	Blakeney Bridges	A48	Gloucestershire	Masonry/stone			Low	Original	Minor culverts	No
1850	Deene Bridge	A43	Northamptonshire	Masonry arch	Y	Fabricated steel beam extension	Medium	Original one side	Minor culvert	No
1850	Desborough Railway Bridge	A6	Northamptonshire	Brick arch			Medium, high from railway	Original	Medium importance rail bridge	No
1850	Disley Station Bridge	A6	Cheshire	Brick arch			High, from platform	Original	Medium importance civic bridge	No
1850	Dunchurch Station North	A45	Warwickshire	3-span Brick arch			Low	Original	Medium importance civic bridge	No
1850	Freeth Canal Bridge	A5	Staffordshire	Brick arch	Y	Simply supported R.C. slab extension on both sides	High from canal and footpaths	Extension both sides	Medium importance civic bridge	No
1850	Freshwater Bridge	A628	Derbyshire	Simply supported R.C. solid slab with masonry piers/abutments			Low	Original	Medium–low importance civic bridge	No
1850	Hauxton Mill Bridge	A10	Cambridgeshire	Brick arch with concrete Infill	Y	In situ R.C. extension	Low	Original one side	Medium	Ancient monument
1850	Heyford Grange No. 1 Bridge	A5	Northamptonshire	Brick arch	Y	Simply supported R.C. slab extension	Medium	Original one side	Medium–low importance civic bridge	No
1850	Marbleflat Bridge	A69	Cumbria	Masonry arch	Y	R.C. arch extension 1937	Low, obscured by vegetation	Original one side	Medium importance civic bridge	No
1850	Massey's Lodge	A49	Cheshire	Cast Iron and Brick Beam and Jack arch		Propped with concrete supports	Can be seen from footpath that runs under bridge	Original now with 3-spans	Medium	No
1850	Nether Tabley	A556	Cheshire	Masonry arch	Y	In situ R.C. box beam and slab extension	Medium/low	Original one side	Medium importance civic bridge	No

Table A.1. Continued

Year	Structure name	Road	County	Original Bridge Form/materials	Extension One side	Extension Both sides	Modifications	Visibility	Visual condition	Importance	Heritage listing
1850	Oakmere Railway Bridge	A556	Cheshire	Masonry and Brick arch				Low-hidden by trees	Original	Medium importance civic bridge	No
1850	Stablecross Railway Bridge	A65	N. Yorkshire	Ashular masonry & Brickwork				Medium	Original	Medium importance rail bridge	No
1850	Temon Bridge	A69	Cumbria	Masonry arch		Y	Repaired 1850, widened 1935	Medium, partially obscured by vegetation	Original bridge hidden by extensions	Minor culvert	No
1850	Thormanby Bridge	A19	N. Yorkshire	Coursed masonry/Brick arch				Low, over disused railway	Original	Low importance civic bridge	No
1850	Tom Otter Bridge	A57	Lincolnshire	Brick arch	Y		Arch re-built 1985, in situ R.C. box extension	Medium	Original one side	Medium importance civic bridge	No
1850	Waithe Beck Bridge	A16	Lincolnshire	Masonry & R.C. twin arch	Y			Medium	Original	Medium importance civic bridge	No
1850	Waverton Railway	A41	Cheshire	Brick arch			R.C. saddle	Medium	Original, but with new saddle	Medium importance rail bridge	No
1851	Private Subway No. 1	A646	Lancashire	Masonry arch	Y		Simply supported concrete widening	Low	Original one side	Low importance civic bridge	No
1851	Private Subway No.2	A646	Lancashire	Masonry arch				Low	Original	Low importance civic bridge	No
1852	Fernhill Heath Railway Bridge	A38		3-span Brick arch			End spans filled in	Medium, not visible from road	Central arch original	Medium importance rail bridge	No
1853	Dover Priory Railway	A20	Kent	3-span brick arch	Y		Steel/concrete continuous beam and slab extension	High from railway platform	Original one side	Medium importance rail bridge	No
1853	Woofferton Skew Rly		Shropshire	Mass fill masonry brick arch				Original from railway	Original	Medium importance rail bridge	No

Year	Name	Road	County	Construction		Alteration	Condition	Originality	Importance	Listed
1856	Bromham Road Railway Bridge	A428	Bedfordshire	2-span Brick arch			High, original from railway	Original	Medium importance civic bridge	No
1856	Knights Bridge	A339	Hampshire	2-span brick/masonry arch	Y	Extended by brick faced R.C.	Medium	Original one side	Medium	No
1857	Burcote Bridge	A43	Northamptonshire	Brick arch	Y	simply supported R.C. extensions each side	Low	Altered both sides, original hidden	Minor culvert	No
1857	Needless Railway	A595	Cumbria	Single-span masonry arch			Low	Original	Medium importance civic bridge	No
1858	Ampthill Road Bridge	A6	Bedfordshire	3-span simply supported beam & slab, W.I. & concrete		strengthened 1968	Medium	Original	Medium importance rail bridge	No
1859	Trethawle Bridge	A38	Cornwall	Masonry arch	Y	Simply supported P.C. extension	Low	Original one side, altered other	Medium–high importance	No
1860	Leadenham (infilled)	A17	Lincolnshire	Brick/stone, filled			Buried under bank Lost		No	No
1860	Wilnecote Station, orig	A5	Staffordshire	Brick arch	Y	Widened	High	Good	Low	No
1860	Newtown Railway	A5	Staffordshire	Brick arch	Y	Widened	Low	Good	Low	No
1860	Waterloo Railway	A565	Merseyside	Fab. steel beam & slab			Medium	Good	Low	No
1860	Riddicks Yard	A6	Derbyshire	Stone culvert			Low	Good	Low	—
1860	Glingerfoot	A7	Cumbria	Masonry arch			Low, hidden by vegetation	Original one side	Medium	No
1860	Kings Hill Bridge (old)	A43	Northamptonshire	Segmental 3-ring brick arch	Y	Widened with blue brick	Low	Patchy, different coloured brickwork	Medium–low importance civic bridge	No
1861	Brockholes Bridge	A59	Lancashire	7-span masonry arch	Y	R.C. widening	High	Original one side	Major civic bridge	No
1861	Llynclys Bridge	A483	Shropshire	Simply supported C.I. girders and brick masonry			Low, hidden by vegetation	Original	Medium–low importance civic bridge	Ancient monument

Table A.1. Continued

Year	Structure name	Road	County	Original Bridge Form/materials	Extension One side	Both sides	Modifications	Visibility	Visual condition	Importance	Heritage listing
1861	Woodlands Railway	A550	Cheshire	Red Sandstone and Brick arch				Low, from railway	Original	Medium importance civic bridge	No
1864	Ampthill Road East Railway Bridge	A6	Bedfordshire	3-span simply supported brick arch beam and Jack arches			Brickwork diaphragms added 1968	Medium	Original	Medium importance rail bridge	No
1865	Redhill Railway Bridge	A49	Herefordshire	3-span brick arch				Original from railway	Original	Medium importance rail bridge	No
1867	Great North Road Railway Bridge No. 109	A638	S. Yorkshire	Red-brick semi-circular skew arch				Medium	Original	Medium importance rail bridge	No
1867	Tidbury Railway Bridge	A303	Hampshire	Brick arch	Y		Concrete encased RSJ simply supported extension	Medium low, fairly concealed	Original one side	Medium–low importance civic bridge	No
1870	Holbeche No 2 (m/arch)	A449	Staffordshire	Brick culvert	Y		Widened	Low	Poor	Low	No
1870	Holbeche No 1 (m/arch)	A449	Staffordshire	Brick culvert	Y		Widened	Low	Poor	Low	No
1870	Calveley Station	A51	Cheshire	Brick, arch				High	Good	Low	No
1870	West Garforth Railway No. 4	A63	West Yorkshire	Brick arch			Deck replacement 1956. Rolled steel beams encased in concrete with *in situ* concrete slab.	Medium, from road below	Original abutments, new deck	Medium importance civic bridge	No
1870	Crowden Brook Bridge	A628	Derbyshire	Random gritstone block single arch				Low	Original	Minor culvert	No
1870	Derby St. Railway Bridge	A570	Lancashire	3-span masonry clay bricks & sandstone block arch				Medium	Original	Medium importance rail bridge	No
1872	Croxdale Railway Bridge	A167	Durham	Brick arch			R.C. saddle added	Medium, highly visible from railway	Original plus saddle	Medium–low importance civic bridge	No

Year	Bridge	Road	County	Structure		Modification	Visibility	Original	Importance	
1874	Victoria Bridge	A30	Cornwall	Masonry/P.C. simply supported beam and slab		Re-decked	Low, only seen from railway	Original supports, new deck	Medium importance civic bridge	No
1875	Coastley Bridge	A69	Northumberland	Masonry arch	Y	R.C. solid slab box wall extension	Medium, obscured by vegetation	Original one side	Medium–low importance civic bridge	No
1875	Lipwood Bridge	A69	Northumberland	Masonry arch	Y	Concrete solid slab arch extension	Medium, obscured by vegetation	Original one side	Medium–low importance civic bridge	No
1877	Great Dudlands Railway Bridge	A59	Lancashire	Simply supported precast & R.C. slab			Medium, highly visible from railway	Original	Medium–low importance civic bridge	No
1878	Bakewell Road culvert	A6	Derbyshire	Brick/stone culvert	Y	Widened	Low	Medium	Low	No
1879	Norbury Bridge	A523	Greater Manchester	Masonry arch		Steel parapets added	Low	Original	Medium importance civic bridge	No
1880	River Bollin culvert	A523	Cheshire	Stone arch	Y	Widened	Medium	Good	High	No
1880	Wood Brook	A6	Leicestershire	Brick, stone, long culvert			Hidden	Not known	Low	No
1880	Highnam Bridge	A48	Gloucestershire	Brick arch			Low	Original	Medium	No
1880	Lansdown Road Bridge	A40	Gloucestershire	Brick and mass concrete arch		Lengthened second span, *in situ* R.C. with brick facing	High, from railway	Original plus 2nd span	Medium importance rail bridge	No
1880	Wansford Railway Bridge	A47	Cambridgeshire	Brick arch			Low, hidden by vegetation	Original	Medium–low importance civic bridge	No
1881	Barmere Bridge	A49	Cheshire	Blue Brick arch			Low	Original	Minor culvert	No
1881	Cholmondeley Bridge	A49	Cheshire	Brick arch			Low	Original	Minor culvert	No
1881	Palmers Bridge	A30	Cornwall	Masonry arch	Y	R.C solid slab/masonry extension	Medium	Original one side	Medium importance civic bridge	No
1881	Quoisley Bridge	A49	Cheshire	Blue Brick arch			Low, hidden by vegetation	Original	Minor culvert	No

Table A.1. Continued

Year	Structure name	Road	County	Original Bridge Form/materials	Extension One side	Both sides	Modifications	Visibility	Visual condition	Importance	Heritage listing
1882	Carlton Station Railway Bridge	A1041	N. Yorkshire	Simply supported wrought iron girders in beam and slab configuration with brick Jack arches				Medium, highly visible from railway	Original	Medium–low importance civic bridge	No
1884	Vale Bridge	A41	Buckinghamshire	Twin-span blue brickwork arch				Medium	Original	Medium importance civic bridge	No
1887	Carminnow Cross Railway Bridge	A38	Cornwall	Brick/masonry arch			Overslabbed, R.C.	Low, hidden by vegetation	Original	Medium–low importance civic bridge	No
1887	Carminnow Cross Slip Road Bridge	A30(S)	Cornwall	Masonry arch	Y		In situ concrete solid slab deck replacement	Low	Original removed and replaced. Masonry abutments exist	Medium importance slip road bridge	No
1887	Onibury Bridge	A49	Shropshire	Wrought iron lattice girders				High, from river	Original	Medium importance civic bridge	No
1889	Leicester Road Railway Bridge	A6	Leicestershire	Simply supported beam and slab			Re-decked	Medium, highly visible from railway	Original plus new deck	Medium importance rail bridge	No
1890	Park Street Railway	A1041	N. Yorkshire	Rolled steel/brick pieces, other forms				From railway	Poor, 1946 new deck	Low	No
1890	Westwell Leacon	A20	Kent	Single-span brick arch	Y		Widened	Minor village	Good	Minor	No
1890	Anchor	A452	West Midlands	Brick, beam & slab				High	Good	Low	No
1891	Beckside Bridge	A595	Cumbria	Masonry arch			Saddled 1992	Low	Original but saddled	Minor culvert	No
1891	South Witham	A1	Lincolnshire	Brick arch		Y	Precast concrete slab and box beams	Medium	Original from east, hidden from west elevation	Major civic bridge	No

Date	Name	Road	County	Structure		Alteration	Assessment	Condition	Importance	Listed
1892	Monkbretton Bridge	A259	East Sussex	R.C. & wrought iron slab on toughing			High	Original	Medium importance civic bridge	No
1893	Limeworks Railway Bridge	A638	S. Yorkshire	4-ring brick arch	Y	R.C. beams added both sides	Low, not visible from main	Original bridge hidden	Medium–low importance civic bridge	No
1893	Newlands Bridge	A590	Cumbria	Simply supported plate girder with steel plates (beam and slab)			Low	Original	Minor culvert	No
1893	Shipton Bridge	A1079	Humberside	Simply supported masonry/RSJ and Jack arches	Y	In situ R.C. slab widened & strengthened	Medium	Original one side	Medium importance civic bridge	No
1886	Dorrington Railway Bridge	A49	Shropshire	Mass fill masonry brick arch			High, from railway	Original	Medium importance rail bridge	Grade 2
1897	Guldeford Lane Corner	A259	East Sussex	Brick culvert	Y	Widened	Hidden	Not known	No	No
1897	Cross Keys Swing Bridge	A17	Lincolnshire	Steel battledeck, brick/concrete Supports		Refurbished 1989	High	Original refurbished	Medium importance swing bridge	No
1899	Gallows No. 1 Bridge	A41	Buckinghamshire	3-ring red brickwork arch	Y	In situ R.C. simply supported extension	Low, partially hidden by vegetation	Original hidden both sides	Minor culvert	No
1900	Dorchester Hill Hatches	A35	Dorset	Brick culvert	Y	Widened	Low	Poor	Low	No
1900		A36	Wiltshire	Twin-span brick culvert	Y	Widened	Low	Not known	Low	No
1900	Burscough Railway	A59	Lancashire	Fab. steel beam & slab			Low	Medium	Low	No
1900	Redlead Mill Bridge	A61	Derbyshire	Twin-span masonry arch	Y		Medium	Original	Low importance civic bridge	No
1900	Beckgrange	A69	Cumbria	Masonry arch	Y	R.C. slab with mass concrete abutments	Low	Original one side	Med/Low importance	No
1900	Brook House Bridge	A446	Warwickshire	Brick arch	Y	Simply supported in situ R.C. & pre-stressed extension	Medium	Original one side	Medium–low importance civic bridge	No
1900	Castle Gardens Bridge	A596	Cumbria	Masonry arch			Low, vegetation	Original	Minor culvert	No

Table A.1. Continued

Year	Structure name	Road	County	Original Bridge Form/materials	Extension One side	Extension Both sides	Modifications	Visibility	Visual condition	Importance	Heritage listing
1900	Cowan Old Bridge	A65	Lancashire	Uncoursed masonry arch				Low, hidden by vegetation	Original	Minor culvert	No
1900	Hungry Hill Bridge	A5	Warwickshire	3-span Brick arch				Medium–low, bridges disused railway	Original	Medium–low importance civic bridge	No
1900	Langley Brook Bridge	A446	Warwickshire	Twin-span brick arch	Y		In situ R.C. extension on one side & R.C. saddle	Medium	Original one side	Medium–low importance civic bridge	No
1900	Micklethwaite Bridge	A596	Cumbria	Masonry arch	Y		In situ R.C. arch added	Low, due to vegetation	Original one side	Minor culvert	No
1900	Old Beck Washdyke Bridge	A52	Lincolnshire	Two ring brick arch, simply supported in situ concrete arch and R.C. saddle				Medium	Original	Minor culvert	No
1900	River Witham	A1	Lincolnshire	2-span brick/masonry arch		Y	In situ R.C. solid slab either side, strengthened 1993	Low	Concrete structure either side	Minor culvert	No
1900	Sheepwash Corner	A628	Derbyshire	3-span masonry arch				Low	Original	Minor culvert	No
1900	South Leeds Railway No. 7 Bridge	M621	W. Yorkshire				Deck replaced 1997	High		Major civic bridge	No
1900	Uffington Road Railway Bridge	A16	Lincolnshire	Brick Jack arches & RSJ			Metal plate parapet added	Low, hidden by vegetation	Original	Medium importance rail bridge	No
1900	Watford Locks Bridge	A5	Northamptonshire	Brick arch	Y		Existing arches saddled and simply supported R.C. slab extension	Low	Original one side	Minor culvert	No
1900	Weedon Railway Bridge	A45	Northamptonshire	Wrought iron and brick, plate girders and Jack arches, simply supported deck				Medium, high from railway		Medium importance rail bridge	No
1900	Wilmington Underbridge	A35	Devon	Masonry arch and 2 minor arches				Medium	Original	Minor culvert	No
1900	Grainsby	A16	Lincolnshire	Brick			Buttressed	Medium	Poor	Medium	No

Year	Name	Road	County	Structure		Modification	Visibility	Originality	Importance	Listed
1900	Woolscott River	A45	Warwickshire	Masonry Twin arch	Y	R.C. solid slab one side, *in situ* R.C. twin box on other side	Low	Original hidden by concrete extensions	Minor culvert	No
1901	Clickmin culvert	A696	Northumberland	*In situ* concrete solid Slab	Y	R.C. solid slab extensions each side	Low	Original hidden by extensions	Low importance minor culvert	No
1902	Caves Inn Bridge	A5	Warwickshire	Brick arch	Y	R.C. simply supported box extensions	Medium	Original hidden both sides by extensions	Medium–low importance civic bridge	No
1902	High Ulverston Bridge	A590	Cumbria	Brick and concrete 3 ring arch			Low, hidden by vegetation	Original	Minor culvert	No
1902	Monk Fryston Bridge	A63	N. Yorkshire	Simply supported composite steel beams and concrete slab, brick abutments		New deck added 1998	High		Major Civic Bridge	No
1903	Aschurch Station Bridge	A438	Gloucestershire	Simply supported steel/brick, plated beams/jack arches bridge			Medium, highly visible from railway	Original	Medium–low importance civic bridge	No
1903	Bankend Bridge	A596	Cumbria	Masonry arch			Low, due to surrounding vegetation	Original	Minor culvert	No
1903	New Bridge	A596	Cumbria	Steel plate girders and buckle plates with concrete infill			High, from road beneath	Original	Medium importance civic bridge	No
1905	Hough Railway Bridge	A500	Cheshire	W.I. girders & brick Jack arches (simply supported)		Raised 1960, overspan 1992	High, particularly from railway		Medium Importance Rail Bridge	No
1906	Cringle Brook	A1	Lincolnshire	Brick arch	Y	*In situ* R.C. solid slab extension in 1960	One side concrete continuous box	Original one side	Minor culvert	No
1906	Galloper Pool Bridge	A595	Cumbria	Engineering brick 3-Ring arch		Structure re-built 1906	Low, hidden by vegetation	Original	Minor culvert	No
1906	Low Byrness Bridge	A696	Northumberland	Masonry solid slab arch	Y	Extended & saddled with R.C. 1957	Low, hidden by vegetation	Original one side	Minor culvert	No

Table A.1. Continued

Year	Structure name	Road	County	Original Bridge Form/materials	Extension One side	Both sides	Modifications	Visibility	Visual condition	Importance	Heritage listing
1906	Pike House Railway Bridge	A4	Avon	Fab steel/in situ mass concrete simply supported beam and slab				Medium, highly visible from railway	Original	Medium importance rail bridge	No
1907	Curdworth Railway Bridge	A446	Warwickshire	5-span brick arch				Medium, highly visible from railway	Original	Medium importance rail bridge	No
1907	Otterburn Bridge	A696	Northumberland	2-span brick arch				Medium, Partially obscured by vegetation	Original	Medium–low importance civic bridge	No
1908	Holderness Drain Bridge	A1033	Humberside	In situ R.C. arch				Medium	Original	Medium–low importance civic bridge	No
1908	Railway No. 1 Bridge, Highfelds	A638	S. Yorkshire	Simply supported fab/steel & blue brick beam and Jack arch bridge				Medium, highly visible from railway	Original	Medium importance rail bridge	No
1909	Barnswood	A523	Staffordshire	Stone arch				Low	Original	Low importance	No
1909	Ryecroft Gate	A523	Staffordshire	Stone arch			New concrete deck (1985)	Low	Original	Minor importance farm access	No
1909	Sanbed Bridge	A646	W. Yorkshire	Simply supported concrete infill between rolled steel beams				Low, small structure	Original	Minor culvert	No
1909	Sewer/Tramway Bridge	M6	West Midlands	In situ R.C. voided box beam				High	Original	Major civic bridge	No
1910	Jogging Bridge	A65	Lancashire	Masonry arch		Y	R.C. simply supported extensions	Medium	Original hidden	Minor culvert	No
1910	Langsett Bridge	A616	S. Yorkshire	In situ concrete arch with masonry face at spandrels				Low, partially hidden by vegetation	Original	Medium–low importance civic bridge	No

Year	Name	Road	County	Type		Extension	Visibility	Condition	Importance	Listed
1910	White Bridge	A41	Merseyside	Brick/Masonry arch			Medium, visible from railway	Original	Medium importance rail bridge	No
1911	Beeston B Bridge	A49	Cheshire	Masonry arch			Medium	Original	Minor culvert	No
1911	Railway No.39 Bridge	A638	S. Yorkshire	Simply supported fab/steel & blue brick beam & Jack arch bridge			Low	Original	Medium importance rail bridge	No
1912	Beeston Canal	A49	Cheshire	In situ R.C. solid slab arch			High, by canal	Original	Medium	No
1912	Belshiel Bridge	A696	Northumberland	Brick arch	Y	R.C. slab extension	Low	Original one side	Minor culvert	No
1912	Coldcoats West Bridge	A696	Northumberland	Brick arch			Medium	Original	Medium–low importance civic bridge	No
1912	Compton	A428	Northamptonshire	Single-span brick arch	Y	Precast R.C. solid slab portal	Medium	Original one side, concrete on other	Medium importance civic bridge	No
1912	Crow Hall west Bridge	A69	Northumberland	Masonry arch with R.C. saddle	Y	Simply supported R.C. solid slab extension	Low	Original one side	Minor culvert	No
1912	Newtown Bridge	A34	Hampshire	3-span brick arch	Y	In situ R.C. extension	Medium	Original one side, other brick faced R.C. arch	Medium importance civic bridge	No
1912	Penwortham New Bridge	A59	Lancashire	3-span masonry arch			Medium–High	Original, possibly a few minor repairs to masonry	Medium, high importance civic bridge	No
1913	Sluice	A565	Lancashire	Rolled steel/beam & slab			High	Good	Low	No
1913	Back drain	A565	Lancashire	Single-span masonry arch			Medium	Original	Medium importance civic bridge	No
1913	Meols Bridge	A565	Lancashire	Masonry and brick arch	Y	R.C. and brickwork extension	Medium	Original one side	Medium importance civic bridge	No

Table A.1. Continued

Year	Structure name	Road	County	Original Bridge Form/materials	Extension One side	Extension Both sides	Modifications	Visibility	Visual condition	Importance	Heritage listing
1913	Pinfold Bridge	A570	Lancashire	In situ R.C. Beam and slab portal				High, from river & footpath	Original	Medium importance civic bridge	No
1913	Sandy Bridge	A565	Lancashire	Masonry and brick arch	Y		R.C. slab extension	Medium, from river	Original one side	Medium importance civic bridge	No
1913	Sluice C	A565	Lancashire	Simply supported fabricated steel beam and floor plate				Medium	New aluminium parapet	Medium importance civic bridge	No
1913	Watergall Under Bridge	A423	Warwickshire	Brick arch	Y		Extended with in situ arched R.C. solid slab	Low	Original one side	Medium–low importance civic bridge	No
1914	Burscough Canal Bridge	A59	Lancashire	In situ R.C. ribbed slab portal			retaining wall added next to bridge	Medium–high	Original	Medium importance civic bridge	No
1914	Whitelee	A68	Northumberland	Masonry arch				Medium	Original	Medium importance civic bridge	No
1914	Shiningpool	A696	Northumberland	Simply supported steel beams/jack arches				Medium	Original	Medium importance civic bridge	No

Index

Page numbers in brackets refer to figures

BRIDGE INDEX

GENERAL INDEX